Interfacing Geostatistics and GIS

Jürgen Pilz (Ed.)

Interfacing Geostatistics and GIS

Springer

Editor
Prof. Dr. Jürgen Pilz
Universität Klagenfurt
Institut für Statistik
Universitätsstr. 65-67
9020 Klagenfurt
Austria
juergen.pilz@uni-klu.ac.at

ISBN 978-3-642-06982-6 e-ISBN 978-3-540-33236-7

DOI 10.1007/978-3-540-33236-7

Cover design: deblik, Berlin

Printed on acid-free paper

9 8 7 6 5 4 3 2 1

springer.com

Preface

Most of the papers contained in this volume grew out of presentations given at the International Workshop StatGIS03 – Interfacing Geostatistics, GIS and Spatial Data Bases, which was held in Pörtschach, Austria, Sept. 29–Oct. 1, 2003, and ensuing discussions, afterwards. Some of the papers are new and have not been given at the conference. Therefore, most of the papers should not be considered as conference proceedings in its original sense but rather more as self-contained and actual contributions to the theme of the conference, the interfacing between geostatistics, geoinformation systems and spatial data base management.

Although some progress has been made toward interfacing, we still feel that there is only little overlap between the different communities. The present volume is intended to provide a bridge between specialists working in different areas. According to the topics of the above mentioned workshop, this volume has been divided into three parts:

Part I starts with general aspects of geostatistical model building (Pebesma) and then new methodological developments in geostatics are presented, in particular this pertains to neural networks (Parkin and Kanevski), Gibbs fields as used in statistical physics (Hristopulos). Furthermore, new developments in Bayesian spatial interpolation with skewed heavy-tailed data and new classification methods based on wavelets (Hofer et al.) and support vector machines (Chaouch et al.) are presented.

Part II contains applications of geostatistics to such diverse areas as geodetic network modelling (Čepek and Pytel), land use policy (Müller and Munroe), precipitation fields modelling (Ahrens), air pollution monitoring (Shibli and Dubois), soil characterization (Sunila and Horttanainen) and soil contamination modelling (Palaseanu-Lovejoy et al.). But also new application areas such as traffic modelling (Braxmeier et al.) and spatial modelling of entrepreneurship data (Breitenecker et al.) are touched.

Part III is devoted to the issues of the integration of different types of information systems. The paper by Krivoruchko and Bivand deals with the problems of interfacing GIS and spatial statistics software systems, from the perspecitves of users and developers. An application of GIS in connection with spatial analysis of remotely sensed agricultural data is reported by Sambrakos and Tsiligiridis.

The advance of the integration efforts with regard to epidemiological information systems is documented in the paper by Gómez-Rubio et al., similar issues arising in biostatistical applications such as acute coronary heart disease are considered by Niyonsenga et al. Non-standard developments and applications of temporal GIS are reported by G. and N. Andrienko, whereas A. Gebhardt reports on severeal possibilities for combining open-source (spatial) databases and GIS.

Finally we would like to thank the Springer Publishing House for offering us the opportunity to publish the present material on the important aspects of integration and combination of spatial modelling branches which previously had developed in a more or less isolated manner. We are looking forward to reporting on further progress in this direction very soon.

Klagenfurt, Spring 2008 *Jürgen Pilz*

Contents

Part II Geostatistical Applications

Part III Integrated Information Systems: Combining (Geo)Statistics, GIS and RDBMS

List of Contributors

Bodo Ahrens
Institut für Meteorologie und Geophysik, Universität Wien, Wien, Austria
Bodo.Ahrens@univie.ac.at

Gennady Andrienko
Fraunhofer Institute AIS Schloss Birlinghoven, Sankt Augustin, Germany
gennady.andrienko@ais.fraunhofer.de

Natalia Andrienko
Fraunhofer Institute AIS Schloss Birlinghoven, Sankt Augustin, Germany

Robert Barr
School of Geography, The University of Manchester, Manchester, UK

Goze Bénié
Geography and Remote-Sensing Department, Université de Sherbrooke, Sherbrooke, QC, Canada
Goze.Bertin.Benie@USherbrooke.ca

Hans Braxmeier
Department of Applied Information Processing, University of Ulm, Ulm, Germany
hans.braxmeier@uni-ulm.de

Robert J. Breitenecker
Department of Innovation Management and Entrepreneurship, University of Klagenfurt, Klagenfurt, Austria
robert.breitenecker@uni-klu.ac.at

Roger Bivand
Norges Handelshøyskole, Bergen, Norway
Roger.Bivand@nhh.no

Aleš Cepek
Faculty of Civil Engineering, CTU Prague, Prague, Czech Republic
cepek@fsv.cvut.cz

A. Chaouch
Institute of Mineralogy and Geochemistry, University of Lausanne, Lausanne, Switzerland
aziz.chaouch@etu.unil.ch

Josiane Courteau
PRIMUS group, Clinical Research Center, Centre Hospitalier Universitaire de Sherbrooke, Sherbrooke, QC, Canada
josiane.courteau@usherbrooke.ca

Charmaine Dean
Statistics and Actuarial Science, Simon-Fraser University, Vancouver, BC, Canada
dean@stat.sfu.ca

Ian Douglas
School of Geography, The University of Manchester, Manchester, UK

Gregoire Dubois
Radioactivity Environmental Monitoring, Institute for the Environment and Sustainability, Joint Research Centre, European Commission, Ispra, Italy
gregoire.dubois@jrc.it

J. Ferrándiz
Dpto. Estadística e Investigación Operativa, Universitat de València, València, Spain
Juan.Ferrandiz@uv.es

Albrecht Gebhardt
Departement of Statistics, University of Klagenfurt, Klagenfurt, Austria
agebhard@uni-klu.ac.at

V. Gómez-Rubio
Dpto. Matemáticas, Universidad de Castilla-La Mancha, Albacete, Spain
Virgilio.Gomez@uclm.es

Thorgeir S. Helgason
Petromodel Ltd, Reykjavik, Iceland
thorgeir@petromodel.is

Abbas Hemiari
PRIMUS group, Clinical Research Center, Centre Hospitalier Universitaire de Sherbrooke, Sherbrooke, QC, Canada

Dionissios T. Hristopulos
Department of Mineral Resources Engineering, Technical University of Crete, Crete, Greece
dionisi@mred.tuc.gr

Vera Hofer
Department of Statistics and Operations Research, Karl-Franzens University Graz, Graz, Austria
vera.hofer@uni-graz.at

Pekka Horttanainen
Department of Surveying, Institute of Cartography and Geoinformatics, Helsinki University of Technology (HUT), Espoo, Finland

Mikhail Kanevski
Institute of Geomatics and Analysis of Risk, University of Lausanne, Switzerland
Mikhail.Kanevski@unil.ch

Hannes Kazianka
Department of Statistics, University of Klagenfurt, Klagenfurt, Austria
hannes.kazianka@uni-klu.ac.at

Konstantin Krivoruchko
Environmental Systems Research Institute Redlands, Redlands, CA, USA
kkrivoruchko@esri.com

A. López
Dpto. Estadística e Investigación Operativa, Universitat de València, València, Spain
Antonio.Lopez@uv.es

M. Maignan
Institute of Mineralogy and Geochemistry, University of Lausanne, Lausanne, Switzerland

Darla K. Munroe
Department of Geography, The Ohio State University, Columbus, OH, USA
munroe.9@osu.edu

Daniel Müller
Leibniz Institute of Agricultural Development in Central and Eastern Europe, Halle (Saale)
mueller@iamo.de

Théophile Niyonsenga
Epidemiology and Biostatistics, Robert Stempel School of Public Health, Florida International University (FIU), Miami, FL (USA)
theophile.niyonsenga@fiu.edu

Monica Palaseanu-Lovejoy
School of Geography, The University of Manchester, Manchester, UK
monica.palaseanu-lovejoy@stud.man.ac.uc

R. Parkin
Institute of Nuclear Safety (IBRAE), Moscow, Russia
park@ibrae.ac.ru

Edzer J. Pebesma
Institute for Geoinformatics (ifgi), University of Münster, Münster, Germany
edzer.pebesma@uni-muenster.de

Jürgen Pilz
Department of Statistics, University of Klagenfurt, Klagenfurt, Austria
juergen.pilz@uni-klu.ac.at

G. Piller
Swiss Federal Office of Public Health (OFSP), Bern, Switzerland

Philipp Pluch
Energy and Petroleum Resources Services GmbH, Vienna, Austria
ppluch@menpet.at

Alexei Pozdnoukhov
Institute of Geomatics and Analysis of Risk, University of Lausanne, Switzerland
Alexei.Pozdnoukhov@unil.ch

Jan Pytel
Faculty of Civil Engineering, CTU Prague, Prague, Czech Republic
pytel@fsv.cvut.cz

J. Rodriguez
Swiss Federal Office of Public Health (OFSP), Bern, Switzerland

M. Sambrakos
InfoLab, Agricultural University of Athens, Athens, Greece
marios@aua.gr

Volker Schmidt
Department of Stochastics, University of Ulm, Ulm, Germany
volker.schmidt@uni-ulm.de

Erich J. Schwarz
Department of Innovation Management and Entrepreneurship, University of Klagenfurt, Klagenfurt, Austria
erich.schwarz@uni-klu.ac.at

Syed Shibli
Landmark Eame Ltd, Aberdeen, Scotland, UK
syed.shibli@googlemail.com

Evgeny Spodarev
Department of Stochastics, University of Ulm, Ulm, Germany
evgeny.spodarev@uni-ulm.de

Gunter Spöck
Department of Statistics, University of Klagenfurt, Klagenfurt, Austria
gunter.spoeck@uni-klu.ac.at

Rangsima Sunila
Department of Surveying, Institute of Cartography and Geoinformatics, Helsinki University of Technology (HUT), Espoo, Finland
rangsima.sunila@hut.fi

T. Tsiligiridis
InfoLab, Agricultural University of Athens, Athens, Greece
tsili@aua.gr

Alain Vanasse
Family Medicine Department, Université de Sherbrooke, Sherbrooke (QC), Canada
alain.vanasse@usherbrooke.ca

Part I

Geostatistical Modeling Aspects and New (Geo)statistical Tools

How We Build Geostatistical Models and Deal with Their Output

Edzer J. Pebesma

Institute for Geoinformatics (ifgi), University of Münster, Münster, Germany
edzer.pebesma@uni-muenster.de

1 Introduction

Multivariable linear geostatistical models extend multivariable, multiple linear regression models for cases where observations are spatially correlated, enabling the prediction of values at unobserved locations. In multiple linear regression, the goal is to explain a large part of the observed variability by a set of regressors and possibly their interactions. The more variability explained, the better the prediction. Geostatistics extends this by looking at spatial correlation in the residual variability: at a prediction location a nearby residual may carry predictive value to the residual value at that location. However, much of the geostatistical curriculum (literature and software) does not start off by attempting to *explain* variability in the observed variables, but rather starts at describing and *modelling* the observed variability *after* assuming the trend is a spatially constant, thereby potentially ignoring available informative predictors.

Extensions are universal kriging and external drift kriging [5]. In universal kriging, only coordinates are used to explain variability. It is of no surprise that this has not become popular, as coordinates hardly ever carry a physical relation to the observed variable, and may lead to extreme, unrealistic extrapolations near the border of the domain. External drift kriging does extend kriging interpolation to the linear using a linear regression model with an external variable for the trend, but it is most often explained as being the case where only a single predictor (external drift variable) is present. In the following, we will not distinguish between universal kriging and external drift kriging, as the procedures are equivalent [7].

Multivariable prediction has been known for a long time, and has been applied especially when using one or more secondary variables to predict a primary variable. The general case where m variables are used to predict m variables, m being larger than say 3, is found seldom in literature. The reasons for this do not have a statistical ground, but rather stem from the fact that it is considered a *burden* to do so. Although algorithms exist to automate the modelling of many direct and cross variograms [10, 11, 24], easily accessible software implementations have been lacking.

This paper discusses several limitations found in statistical software (mostly R) and Geographic Information Systems (GIS; [4]) software, with respect to flexible modelling of the multivariable linear geostatistical model. In a case study we show how to apply this model, using the recently developed `gstat` package for R and S-PLUS [20]. The discussion closes with a highly personalized view on the practice, limitations and possibilities of applied geostatistics.

2 Geostatistical Prediction

In geostatistics, the variability in an observed variable Z, taken at location s_i, is assumed to be the sum of a fixed trend and a random residual: $Z(s_i) = m(s_i) + e(s_i)$, and the trend is modelled as a linear combination of p unknown coefficients and p known predictors $X_j(s)$:

$$Z(s) = \sum_{j=1}^{p} X_j(s)\beta_p + e(s) = X(s)\beta + e(s), \quad s \in \{s_1, ...s_n\}$$

with $X_1(s) \equiv 1$ when β_1 is the intercept, and $X(s)$ the $n \times p$ matrix with predictors. Given knowledge of the (spatial) covariance of e, $V = \text{Cov}(e)$, and knowledge of the covariance between $e(s)$ and $e(s_0)$, $v = (\text{Cov}(e(s_1), e(s_0)), ..., \text{Cov}(e(s_n), e(s_0)))'$, the best linear unbiased (or kriging) predictor is obtained by

$$\hat{Z}(s_0) = x(s_0)\hat{\beta} + v'V^{-1}(Z(s) - X(s)\hat{\beta})$$

where $x(s_0)$ contains the known predictors location s_0, and with

$$\hat{\beta} = (\hat{\beta}_0, ..., \hat{\beta}_1)' = (X(s)'V^{-1}X(s))^{-1}X(s)'V^{-1}Z(s)$$

the generalized least squares estimate of β. The prediction variance of $\hat{Z}(s_0)$ is

$$\sigma^2(s_0) = \sigma_0^2 - v'V^{-1}v + \eta'(X(s)'V^{-1}X(s))^{-1}\eta$$

with $\sigma_0^2 = \text{Var}(e(s_0))$ and $\eta' = (x(s_0) - v'V^{-1}X(s))$. These equations reduce to traditional multiple regression prediction if $v = 0$ and V is diagonal (weighted least squares) or $V = \sigma_0^2 I$ (ordinary least squares) [9], and they reduce to ordinary kriging if the regression only contains an intercept (i.e., $X(s)$ and $x(s_0)$ only contain a single column of ones).

When multiple, spatially cross correlated variables are present, they may be used in a multivariable prediction [23], not only to enhance the predictions of each individual variable, but also to assess the prediction error covariances for all pairs of variables.

In practice, the application of these equations is often restricted to the data available in a local neighbourhood around s_0. The reasons for this may be computational, to avoid solving kriging systems with a very large ($n \gg 1000$) covariance matrix, or statistical, to reduce the assumption of globally constant regression coefficients to the more flexible assumption of locally constant regression coefficients.

Another specialty on the geostatistics menu is called *change of support*: rather than predicting values $Z(s_0)$ for *point* locations s_0, we may want to predict the integral (mean) of $Z(B_0) = \frac{1}{|B_0|}\int_{u \in B_0} Z(u)du$, with $|B_0|$ the area or volume of integration. Block average values can be obtained by averaging point kriging values, but block average prediction errors can not; for this we need block kriging [5, 13]. The reason for wanting block kriging is that highly detailed spatial predictions may not be wanted, and that block kriging prediction errors are always smaller then point kriging prediction errors.

In addition to prediction, it may be useful to *simulate* realisations of random fields $Z(s)$ that honour both the observed data, the regression relations, and the spatial correlation [19]. Abrahamsen and Espen Benth [2] describe an algorithm where the simulation equivalent of universal (external drift) kriging is given.

3 Case Study: Sea Floor Sediment Pollution

This case study analyses spatio-temporal data on sea floor sediment pollution, collected from 1986 to 2000 in the Dutch part of the North Sea [17]. The data set was provided by the Dutch National institute for Coastal and Marine Management (RIKZ). The variables measured comprise heavy metals, as well as organic compounds like furans and polychlorinated biphenyls (PCB's). Here, we will look into a single PCB, named PCB138. Table 1 summarizes some of the characteristics of the data. The programme initially aimed at monitoring approximately five-yearly (the "main" monitoring years, '86, '91, '96, '00), over which the number of samples range from 31 to 49; other measurements result from additional sampling programs.

The main PCB138 source is sediment originating from the river Rhine; this sediment is carried North bound along the coast by the local North Sea flow direction. One of the main questions is whether temporal trends are present in the data, how large the trend is, and to which extent spatially differentiated trends can be inferred.

Table 1. PCB138 (µg/kg dry matter) data summaries; years marked with a * are the regular monitoring years, other years result from additional sampling programs

year	1986*	1987	1989	1991*	1993	1996*	2000*	All
mean	7.29	8.39	4.08	3.70	1.03	1.58	1.27	4.20
median	6.90	7.50	2.65	3.05	0.775	1.40	0.90	2.85
max	21.1	19.7	12.3	13.1	2.7	4.9	3.3	21.1
min	1.60	2.10	1.00	0.70	0.25	0.20	0.20	0.2
n	45	29	14	42	6	49	31	216

3.1 Exploratory Data Analysis

Figure 1 shows a bubble plot with the spatial locations of the measurement sites, per year. Symbol size is proportional to log-concentration, which is the natural scale to view such variables. The summary statistics of Table 1 already reveal that PCB138 decreases over time. Figure 1 furthermore shows that high concentrations appear close to the coast. Simply looking at how PCB138

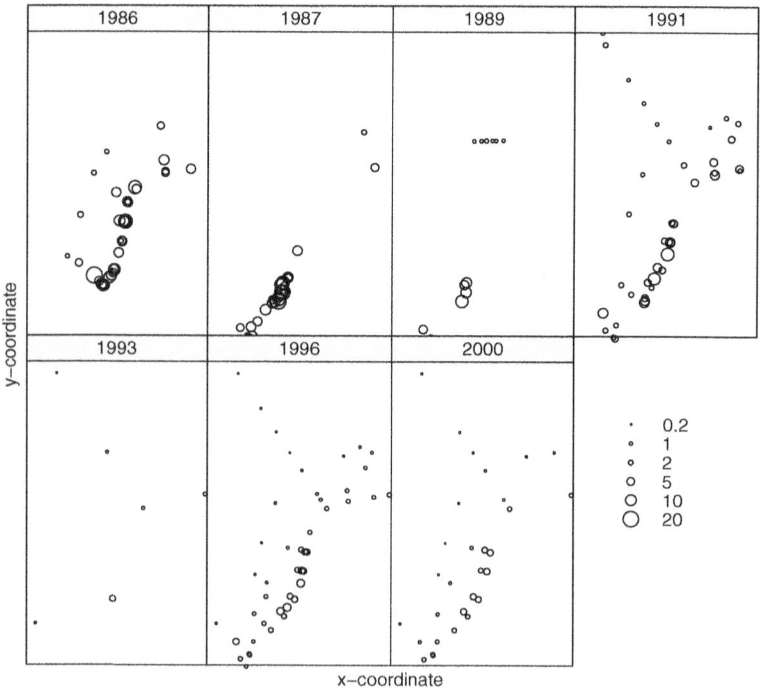

Fig. 1. Maps of measured PCB138 concentrations in Dutch North Sea sediment samples, for each year sampled from 1986 to 2000. The area shown has UTM31 x-coordinates ranging from 464000 m. to 739000 m. and y-coordinates from 5696500 m. to 6131500 m.

concentrations decrease with time may not be appropriate because the spatial locations of sampling vary from year to year, and the sampling pattern is not random. The sampling pattern (Fig. 1) is directed towards transects perpendicular to the Dutch coast (the direction of the main gradient), and seems clustered; many short distances are present.

3.2 Trend

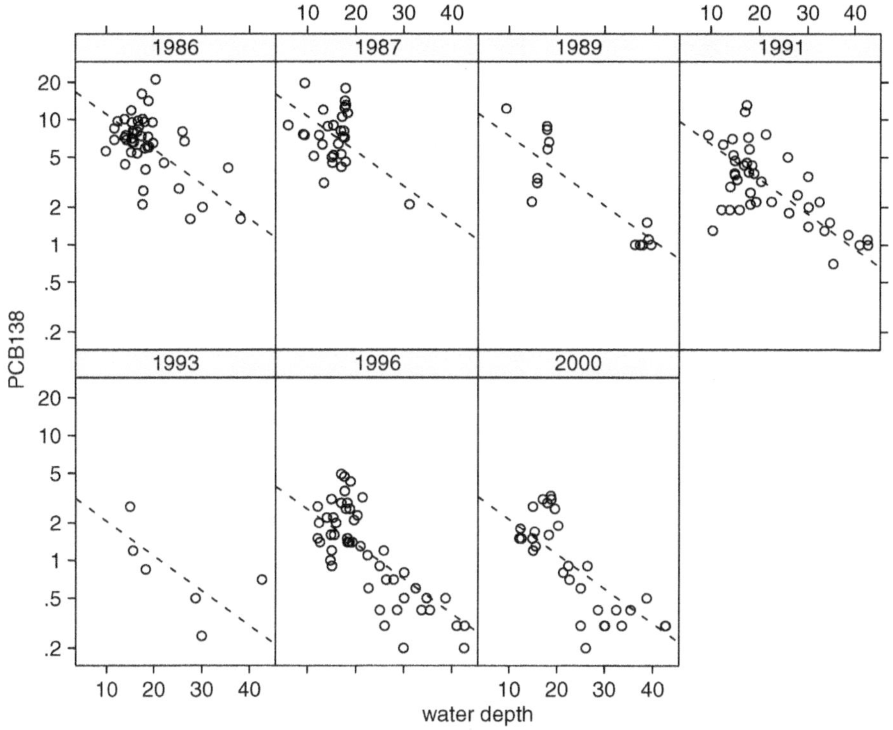

Fig. 2. PCB138 concentration as a function of sea water depth, for each of the measured years

Figure 2 shows spatial trends of (log) PCB138 concentrations as a function of water depth, for each monitoring year. The figure suggests that on the log-scale, the decrease with depth is more or less constant, and that the origin (intercept) of the varies per year; the fitted line reflects this model. In terms of a regression model, we could express this as

$$Z_t(s) = \beta_{0,t} + \beta_1 + X_1(s) + e(s) \tag{1}$$

with $Z_t(s)$ the log-transformed PCB138 measurement at year t and location s, $\beta_{0,t}$ the intercept for year t, with $X_1(s)$ the depth value at location s and

$e(s)$ the residual. The regression model explains 77% of the variability in log-PCB138. Under the assumption of independent data, (i) all terms were highly significant ($p < .001$), and (ii) an interaction between year and depth (i.e., a year-dependent regression slope with depth) was not significant. Clearly, these significance assertions are of little value, as the data vary spatially, and we may assume that they are spatially correlated.

3.3 Residual Spatial Correlation and Temporal Cross Correlation

Each of the monitoring years years has too few measurements to model a residual variogram (Table 1). For that reason, the residual information of all years was merged. Simply merging all residuals leads to the variogram in the first panel of Fig. 3. This would be a valid approach if the residual spatial *pattern* were constant over time. Constructing a pooled variogram by only considering point pairs with both measurements in the same year (rest of Fig. 3) shows that the hypothesis of a temporary constant spatial pattern is not valid: a much stronger spatial correlation is revealed under the hypothesis that only the spatial variability (variogram) is persistent over time. On single residual variogram model was fitted for all within-year residuals, $\gamma(h) = 0.08\delta(h + 0.224(1 - \exp(-h/17247))$ with $\delta(h) = 0$ if $h = 0$ and $\delta(h) = 1$ if $h > 0$ (last panel of Fig. 3).

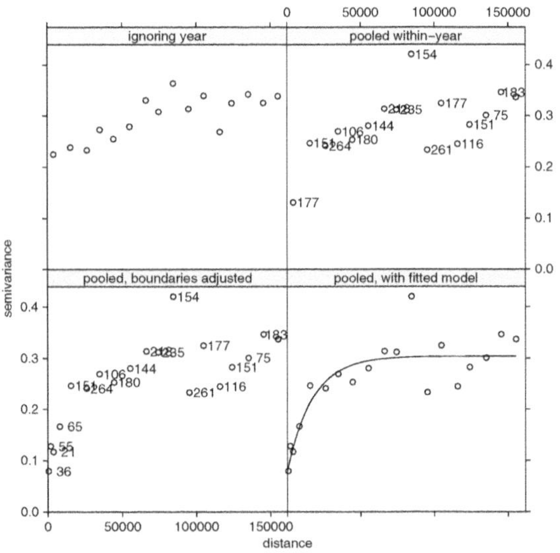

Fig. 3. Different approaches to modelling the sampling variogram for the residuals of the linear regression lines in Fig. 2; *top left*: residual variogram; *top right*: pooled, within-year residual variogram; *bottom left*: short-distance variogram values are split into smaller distance intervals; *bottom right*: a model fitted to the *bottom left* variogram. Numbers reflect the point pairs that contribute to sample variogram estimates

Before looking at residual cross correlation and temporal change, we will restrict ourselves to the four "main" monitoring years, 1986, 1991, 1996 and 2000, because the other years have too few measurements for analyzing cross variograms. The direct variograms for years in Fig. 4 (labeled "1986", "1991",...) show indeed that each of them carry insufficient information for fitting a separate model, and the fitted model used is the same for each, i.e. that of Fig. 3. The cross variograms for pairs of years (labeled "1986.1991", etc.) are even more noisy, and attempts to automatically fit a linear model of coregionalization [11] failed.

In order to proceed, we need a coregionalization model that requires less parameters, such as the intrinsic correlation model [10]. This model only requires a correlation between two years. Finding these correlations from the cross variograms seems not so easy, partly because pairs of years do not share common sample locations (Fig. 1). Therefore, we used the following approach: for a pair of years, say 1986 and 1991, for each of the measurements in 1986 the (spatially) nearest measurement of 1991 was found, and from the pairs

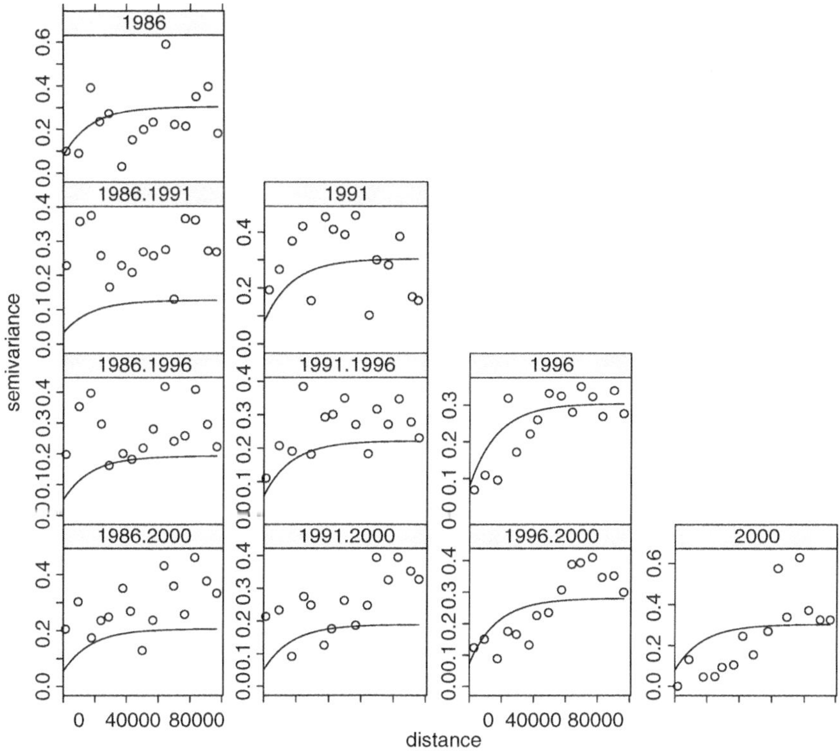

Fig. 4. Sample direct and cross variograms for the four main measurement years, and fitted Intrinsic Correlation model. The direct variogram model is that of 3, each cross variogram is scaled down by a factor equal to the pointwise correlations of the pair of years; pointwise correlations are approximated by joining spatially nearest neighbours to form data pairs

thus found, the correlation coefficient was calculated. Next, the two years were reversed, and a second correlation coefficient was calculated. The average of these two correlations was used to model the cross variograms of Fig. 4. Because spatially nearest neighbours of year y were used to approximate the measured value at a certain location in year x, the estimated correlations must underestimate the true correlations.

3.4 Spatio-Temporal Prediction

Spatio-temporal prediction under model (1), given the data for each of the four "main" years and given the direct variograms and the cross variograms of Fig. 4 is simply a matter of universal cokriging. Universal cokriging yields spatial predictions for each of the four years, shown in Fig. 5, and yields in addition spatial prediction error variances for each of the four years, and spatial prediction error covariances for prediction errors of all pairs of years. Spatially differentiated estimates of trends can be assessed by combining the yearly predictions and prediction error (co)variances.

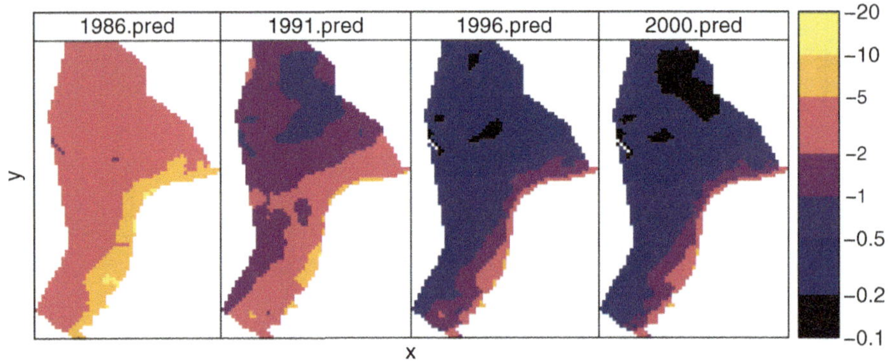

Fig. 5. Cokriging predictions for the four main measurement years

3.5 Contrasts and Trends

Cokriging basically yields for each location s_0 a vector with predictions, which in our case could be called $\mathbf{y}(s_0) = (y_{86}(s_0), y_{91}(s_0), y_{96}(s_0), y_{00}(s_0))'$, along with the prediction error covariance matrix $\mathrm{Cov}(\mathbf{y}(s_0))$. Given this vector we can calculate for each location s_0 a *contrast*

$$C(s_0) = \lambda'\mathbf{y}(s_0)$$

which has prediction error variance $\lambda'\mathrm{Cov}(\mathbf{y}(s_0))\lambda$. Examples of potentially interesting contrasts are

- prediction for a single year, e.g. 1991: $\lambda' = (0, 1, 0, 0)$
- prediction of the four-year mean: $\lambda' = (\frac{1}{4}, \frac{1}{4}, \frac{1}{4}, \frac{1}{4})$

- prediction of the *difference* between the means of 1986 and 1991 versus the mean of 1996 and 2000: $\lambda' = (-\frac{1}{2}, -\frac{1}{2}, \frac{1}{2}, \frac{1}{2})$
- prediction of the average yearly increase: $\lambda' = (-0.065, -0.02, 0.025, 0.061)$

The weights of the latter contrast, which is obviously of major interest when we want to assess spatially differentiated trends, are obtained as follows. Trend estimation uses linear regression for predicting concentrations from years by $\mathbf{y}(s_0) = \beta_0(s_0) + \beta_1(s_0)t + e = X\beta(s_0) + e$. The ordinary least squares estimate of β is $(X'X)^{-1}X'\mathbf{y}$. The contrast coefficients that estimate $\beta_1(s_0)$ are in the second row of $(X'X)^{-1}X'$, with

$$X = \begin{bmatrix} 1 & 1986 \\ 1 & 1991 \\ 1 & 1996 \\ 1 & 2000 \end{bmatrix}.$$

Figure 6 shows the predicted trends, as well as the trend predictions divided by their own prediction standard error. Clearly, the majority of the area

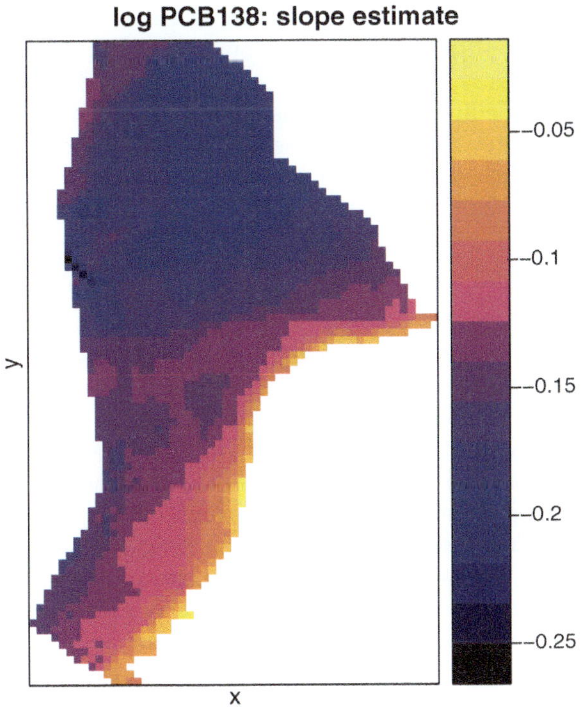

Fig. 6. Predicted trends (in ppm/year) for each point location (*left*); and relative predicted trends, expressed as fraction of their own prediction standard error (*right*). On the right, under the assumed model, relative predicted trends smaller than -2 tentatively indicate trends that cannot be attributed to pure chance (i.e. that are significant)

shows a decrease in PCB138 concentration of a magnitude larger than twice its standard error, tentatively indicating significant trends. Of course, these assertions are only valid under the assumptions that were made along the way, including (i) the model for the trend, shown in Fig. 2, (ii) the modeled variograms and cross variograms (Figs. 3 and 4), and (iii) that on the log-scale, a second order stationary residual for PCB138 is a reasonable model. Stronger positive correlations between the years result in smaller standard errors for trends, so the underestimation of true between-year (spatial) correlations that we mentioned in Sect. 3.3 results in conservative assessment of "significance" of trends in the right part of Fig. 6.

4 Shortcomings

This case study shows some of the capabilities of the `gstat` package for R [12], or for S-PLUS, which extensively uses the graphics capabilities of the Trellis/lattice graphics package [6]. The `gstat` program [18] or R package [20] offers flexibility with respect to trend modelling, multivariate variogram modelling, multivariate prediction and simulation, change of support and prediction in a local neighbourhood. Additional features that it does not address are e.g. flexible three-dimensional anisotropic variogram modelling, Bayesian handling of uncertainty in variogram model coefficients [22], and multivariable space time modelling in continuous time (i.e., where time is a dimension rather than a discrete variable as in the case study of this paper). These features are available in either other R packages or other environments, where they are potentially hard to combine with the features offered by `gstat`.

5 Discussion

There may be various historical reasons for not starting off with a linear regression model for the modelling the trend of spatial data. First, geostatistics was developed by mining engineers, who usually did not have useful predictor variables available, other than spatial coordinates of observations and prediction locations. Second, the sample variogram from estimated residuals is biased because of the estimation of the trend, which would need the true variogram—a chicken-and-egg problem raised by Armstrong [1] but settled by Kitanidis [15]. Third, leading authors have suggested that predictors are not needed [14], but that observations themselves carry enough information. This is indeed the case when observations are abundant and not too noisy—the case where even (geo)statistics could be ignored altogether and any contouring algorithm would suffice. All these factors lead to situation where much of the available geostatistical software packages (GSLIB, [8]; GsTL, [21]; ArcGIS Geostatistical Analyst, [16]; Isatis, `http://www.geovariances.fr`) have little flexibility with respect to modelling external drifts with multiple linear regression models.

In the authors opinion, R (or S-PLUS) provides a very rich environment for building geostatistical models, mainly because all basic building blocks are there: linear and non-linear regression functions, regression diagnostic plotting functions, and the Trellis graphics [6] tools—which we consider indispensable for multivariable geostatistical exploration, analysis, modelling and prediction. In addition, functions are present that allow straightforward calculation of contrasts, or of aggregating quantities over large sets of Monte Carlo simulations; these cases are usually much harder to deal with in a GIS environment. A drawback or R (and S-PLUS) is that it still has no standard way of dealing with spatial data, although several authors now work together in this direction [3]. It will be a long way before R can touch upon commercial GIS' capabilities regarding visualisation of maps, but the basic building blocks are there. In terms of analytical capabilities R has beaten commercial GIS by far.

References

1. Armstrong, M. (1984) Improving the estimation and modelling of the variogram. In: G. Verly et al, (Eds.), Geostatistics for Natural Resource Characterization, pp 1–20, Reidel, Dordrecht, Holland.
2. Abrahamsen, P. and F.E. Benth (2001) Kriging with inequality constraints. Mathematical Geology 33 (6), 719–744.
3. Bivand, R.S. (2003) Approaches to classes for spatial data in R. In: K. Hornik and F. Leisch (Eds.), Proceedings of the 3rd International Workshop on Distributed Statistical Computing (DSC 2003) March 20–22, Vienna, Austria. `http://www.ci.tuwien.ac.at/Conferences/DSC-2003/`
4. Burrough, P.A. and R.A. McDonnell (1998) Principles of geographical information systems. Oxford University Press, New York. 431 pp.
5. Chilès, J-P. and P. Delfiner (1999) Geostatistics, modeling spatial uncertainty. Wiley, New York.
6. Cleveland, W.S. (1993) Visualizing data. Hobart Press, Summit, New Jersey.
7. Cressie, N.A.C. (1993) Statistics for Spatial Data, revised edition. Wiley, New York.
8. Deutsch, C.V. and A.G. Journel (1990) GSLIB geostatistical software library and user's guide, second edition. Oxford University Press, New York.
9. Draper, N.R. and H. Smith (1981) Applied regression analysis, second edition. Wiley, New York.
10. Goovaerts, P. (1997) Geostatistics for natural resources evaluation. Oxford University Press, New York.
11. Goulard, M. and M. Voltz (1992) Linear coregionalization model: tools for estimation and choice of cross-variogram matrix. Mathematical Geology 24 (3), 269–286.
12. Ihaka, R. and R. Gentleman (1996) R: a language for data analysis and graphics. Journal of Computational and Graphical Statistics 5(3) 299–314.
13. Journel, A.G. and Ch.J. Huijbregts (1978) Mining geostatistics. Academic Press, London, 600 pp.
14. Journel, A.G. and M.E. Rossi (1989) When do we need a trend model in kriging? Mathematical Geology 21 (7), 715–739.
15. Kitanidis, P.K. (1993) Generalized covariance functions in estimation. Mathematical Geology 25 (5), 525–540.

16. Krivoruchko, K. (2007)...., this volume.
17. Laane, R.W.P.M., H.L.A. Sonneveldt, A.J. Van der Weyden, J.P.G. Loch and G. Groeneveld (1999) Trends in the spatial and temporal distribution of metals (Cd, Cu, Zn and Pb) and organic compounds (PCBs and PAHs) in Dutch coastal zone sediments from 1981 to 1996: a model case study for Cd and PCBs. Journal of Sea Research 41, 1–17.
18. Pebesma, E.J. and C.G. Wesseling (1998) Gstat, a program for geostatistical modelling, prediction and simulation. Computers & Geosciences, 24 (1), 17–31.
19. Pebesma, E.J. and G.B.M. Heuvelink (1999) Latin hypercube sampling of Gaussian random fields. Technometrics 41 (4), 303–312.
20. Pebesma, E.J. (2004) Multivariable geostatistics in S: the gstat package. Computers & Geosciences 30, 683–691.
21. Remy, N. (2001) GsTL: the geostatistical template library in C++. MSc thesis, Dept. of Petroleum Engineering, Stanford University.
22. Stein, M.L. (1999) Interpolation of spatial data: some theory for kriging. Springer, New York, 247 pp.
23. Ver Hoef, J.M. and N.A.C. Cressie (1993) Multivariable spatial prediction. Mathematical Geology, 25 (2), pp. 219–240.
24. Yao, T. and A.G. Journel (1997) Automatic modeling of (cross) covariance tables using fast fourier transform. Mathematical Geology 30, 589–616.

Spartan Random Fields: Smoothness Properties of Gaussian Densities and Definition of Certain Non-Gaussian Models

Dionissios T. Hristopulos

Department of Mineral Resources Engineering, Technical University of Crete, Crete, Greece
dionisi@mred.tuc.gr

1 Introduction

Spartan spatial random fields (SSRFs) were introduced in [10]. Certain mathematical properties of SSRFs were presented, inference of the model parameters from synthetic samples was investigated [10], and methods for the unconditional simulation of SSRFs were developed [11]. This research has focused on the fluctuation component of the spatial variability, which is assumed to be statistically homogeneous (stationary) and normally distributed. The probability density function (pdf) of Spartan fields is determined from an energy functional $H[X_\lambda(\mathbf{s})]$, according to the familiar in statistical physics expression for the Gibbs distribution

$$f_{\mathrm{x}}[X_\lambda(\mathbf{s})] = Z^{-1} \exp\{-H[X_\lambda(\mathbf{s})]\}\,. \tag{1}$$

The constant Z (called partition function) is the pdf normalization factor obtained by integrating $\exp(-H)$ over all degrees of freedom (i.e. states of the SSRF). The subscript λ denotes the fluctuation resolution scale. The energy functional determines the spatial variability by means of interactions between neighboring locations. One can express the multivariate Gaussian pdf, typically used in classical geostatistics, in terms of the following energy functional

$$H[X_\lambda(\mathbf{s})]=\frac{1}{2}\int d\mathbf{s}\int d\mathbf{s}'\, X_\lambda(\mathbf{s})\, c_X^{-1}(\mathbf{s},\mathbf{s}')\, X_\lambda(\mathbf{s}'), \tag{2}$$

where $c_X(\mathbf{s},\mathbf{s}')$ is the centered covariance function; the latter needs to be determined from the data for all pairs of points $\mathbf{s}$ and $\mathbf{s}'$, or (assuming statistical homogeneity) for all distance vectors $\mathbf{s}-\mathbf{s}'$. In contrast, the energy functional in Spartan models is determined from physically motivated interactions between neighbors. The name 'Spartan' emphasizes that the number N_p of model parameters to be determined from the data is small. For example, in the fluctuation – gradient – curvature (FGC) model, the pdf involves three main parameters: the scale factor η_0 , the covariance shape parameter η_1, and the correlation length ξ . Another factor that adds flexibility to the model is the coarse-graining kernel that determines the fluctuation resolution λ [10]. As we show below, the resolution is directly related to smoothness properties of the SSRF. In previous work [10, 11], we have used a kernel with a boxcar spectral density that imposes a sharp cutoff in frequency (wavevector) space at $k_c \propto \lambda^{-1}$. We have treated the cutoff frequency as a constant, but it is also possible to consider it as an additional model parameter, in which case $N_p = 4$.

A practical implication of an interaction-based energy functional is that the parameters of the model follow from simple sample constraints that do not require the full calculation of two-point functions (e.g., correlation function, variogram). This feature permits fast computation of the model parameters. In addition, for general spatial distributions (e.g., irregular distribution of sampling points, anisotropic spatial dependence with unknown a priori principal directions), the parameter inference does not require various empirical assumptions such as choice of lag classes, number of pairs per class, lag and angle tolerance, etc. [7] used in the calculation of two-point functions. In the case of SSRFs that model data distributed on irregular supports, the definition of the interaction between 'near neighbors' is not uniquely defined. Determining the neighbor structure for irregular supports increases the computational effort [10], but the model inference process is still quite fast. Methods for the non-constrained simulation of SSRFs with Gaussian probability densities on the square lattice (by filtering Gaussian random variables in Fourier space and reconstructing the state in real space with the inverse FFT) and for irregular supports (based on a random phase superposition of cosine modes with frequency distribution modeled on the covariance spectral density), have been presented in [11].

2 FGC Energy Functional

The energy functional involves the SSRF states (configurations) $X_\lambda(\mathbf{s})$. For notational simplicity, we will not use different symbols for the random field and its states in the following. As hinted above, the energy functional is properly defined for SSRFs $X_\lambda(\mathbf{s})$ with an inherent scale parameter 'λ' that denotes the spatial resolution of the fluctuations. At lower scales, the fluctuations are coarse-grained. The fluctuation resolution scale is physically meaningful, since it would be unreasonable to expect a model of fluctuations to be valid for all length scales. In contrast with classical random field representations, which do not have a built-in scale for a fluctuation cutoff, SSRFs provide an explicit 'handle' for this meaningful parameter. In practical situations, the fluctuation resolution scale is linked to the measurement support scale and the sampling density. In the case of numerical simulations, the lattice spacing provides a lower bound for λ. The fluctuation resolution can also exceed the lattice spacing, to allow for smoother variations of the field. The general probability density function of continuum FGC Spartan random fields (FGC-SSRF) in $\mathbb{R}^d$ is determined from the following functional

$$H_{\mathrm{fgc}}[X_\lambda] = \frac{1}{2\eta_0 \xi^d} \int d\mathbf{s}\, h_{\mathrm{fgc}}\left[X_\lambda(\mathbf{s}); \eta_1, \xi\right], \tag{3}$$

where η_0 is a scale factor with dimensions $[X]^2$ that determines the magnitude of the overall variability of the SSRF, η_1 is a covariance shape parameter (dimensionless), ξ is the correlation length, and h_{fgc} is the normalized (to $\eta_0 = 1$) local energy at the point $\mathbf{s}$. In the case of a Gaussian FGC random field with mean (not necessarily stationary) $m_{X;\lambda}(\mathbf{s}) = E\left[X_\lambda(\mathbf{s})\right]$ and isotropic spatial dependence of the fluctuations, the functional $h_{\mathrm{fgc}}\left[X_\lambda(\mathbf{s}); \eta_1, \xi\right]$ is given by the following

$$h_{\mathrm{fgc}}\left[X_\lambda(\mathbf{s}); \eta_1, \xi\right] = \left[\chi_\lambda(\mathbf{s})\right]^2 + \eta_1\, \xi^2 \left[\nabla \chi_\lambda(\mathbf{s})\right]^2 + \xi^4 \left[\nabla^2 \chi_\lambda(\mathbf{s})\right]^2, \tag{4}$$

where $\chi_\lambda(\mathbf{s})$ is the local fluctuation field. The functional (4) is permissible if Bochner's theorem [3] for the covariance function is satisfied. As shown in [10], permissibility requires $\eta_1 > -2$. The covariance spectral density follows from the equation

$$\tilde{G}_{\mathrm{x};\lambda}(\mathbf{k}) = \frac{\left|\tilde{Q}_\lambda(\mathbf{k})\right|^2 \eta_0\, \xi^d}{1 + \eta_1\, (k\,\xi)^2 + (k\,\xi)^4} \tag{5}$$

where $\tilde{Q}_\lambda(\mathbf{k})$ is the Fourier transform of the smoothing kernel. If the latter is the boxcar filter with cutoff at k_c, (5) leads to a band-limited spectral density $\tilde{G}_{\mathrm{x};\lambda}(\mathbf{k})$. For negative values of η_1 the spectral density develops a sharp peak, and as η_1 approaches the permissibility boundary value equal to -2, the spectral density tends to become singular. For negative values of η_1 the structure of the spectral density leads to a negative hole in the covariance function in real space. If $\tilde{Q}_\lambda(\mathbf{k})$ has no directional dependence, the spectral density depends on the magnitude but not the direction of the frequency vector $\mathbf{k}$. Thus, the covariance is an isotropic function of distance in this case.

On regular lattices, the FGC spectral density is obtained by replacing the operators ∇ and ∇^2 in the energy functional with the corresponding finite differences. Then, the local energy becomes $h_{\mathrm{fgc}}\left[X_\lambda(\mathbf{s}); \eta_1, \xi\right] = h_{\mathrm{fgc}}\left[\chi_\lambda\left\{U(\mathbf{s}); \eta_1, \xi\right\}\right]$, where $U(\mathbf{s}) = \mathbf{s} \cup nnb(\mathbf{s})$ is the local neighborhood set that contains the point $\mathbf{s}$ and its nearest lattice neighbors, $\chi_\lambda\left\{U(\mathbf{s})\right\}$ is the set of the SSRF values at the points in $U(\mathbf{s})$, and $h_{\mathrm{fgc}}[\cdot]$ is a quadratic functional of the SSRF states that defines interactions between the fluctuation values $\chi_\lambda\left\{U(\mathbf{s})\right\}$. For irregular spatial distributions, there are more than one possibilities for modeling the interactions. One approach, explored in [10], is to define a background lattice that covers the area of interest and to construct interactions between the cells of the background lattice. If $C_B(\mathbf{s})$ denotes the cell of the background lattice that includes the point $\mathbf{s}$ and $nnb\{C_B(\mathbf{s})\}$ is the set of nearest neighbors of the cell $C_B(\mathbf{s})$, the local neighborhood set involves the sampled points that belong to the cell $C_B(\mathbf{s})$ and its neighbors, i.e. $U(\mathbf{s}) = \mathbf{s}' \in C_B(\mathbf{s}) \cup nnb\{C_B(\mathbf{s})\}$.

3 Model Inference

The problem of model inference from available data is a typical inverse problem. In order to determine the model parameters experimental constraints need to be defined that capture the main features of the spatial variability in the data. These constraints should then be related to the interactions in the SSRF energy functional. The experimental constraints used in [10] for the square lattice are motivated by the local 'fluctuation energy measures' $S_0(\mathbf{s}) = \chi_\lambda^2(\mathbf{s})$, $S_1(\mathbf{s}) = \sum_{i=1}^{d} \left[\nabla_i \chi_\lambda(\mathbf{s})\right]^2$, and $S_2(\mathbf{s}) = \sum_{i,j=1}^{d} \Delta_2^{(i)} \left[\chi_\lambda(\mathbf{s})\right] \Delta_2^{(j)} \left[\chi_\lambda(\mathbf{s})\right]$, where $\Delta_2^{(i)}$ denotes the centered second-order difference operator. The respective experimental constraints are then given by $\overline{S_0(\mathbf{s})}$ (sample variance), $\overline{S_1(\mathbf{s})}$ (average square gradient) and $\overline{S_2(\mathbf{s})}$, where the bar denotes the sample average. The respective stochastic constraints are $E\left[S_m(\mathbf{s})\right]$, $m = 0, 1, 2$ and they can be expressed in terms of the covariance function. For the isotropic FGC model, calculation of the stochastic constraints involves a one-dimensional numerical integration over the magnitude of the frequency. Matching of the stochastic and experimental constraints is formulated as an optimization problem in terms of a functional that measures the distance between the two sets [10] of constraints. Minimization of the distance functional leads to a set of optimal values $\eta_0^*, \eta_1^*, \xi^*$ for the model parameters. Use of k_c as a fourth parameter needs further investigation. It should be noted that constraint matching is based on the ergodic assumption, and thus a working approximation of ergodicity should be established for the fluctuation field.

4 Smoothness of FGC Spartan Random Fields

The probability density of the FGC-SSRF involves the first- and second-order derivatives of the field's states. This requires defining the energy functional in a manner consistent with the existence of the derivatives. In general, for Gaussian random fields [1, 15], the nth-order derivative $\partial^n X_\lambda(\mathbf{s})/\partial s_1^{n_1} ... \partial s_d^{n_d}$ exists in the mean square sense if (i) the mean function $m_{X;\lambda}(\mathbf{s})$ is differentiable, and (ii) the following derivative of the covariance function exists [1, 15]

$$\left. \frac{\partial^{2n} G_{\mathrm{x};\lambda}(\mathbf{s}, \mathbf{p})}{\partial s_1^{n_1} ... \partial s_d^{n_d} \, \partial p_1^{n_1} ... \partial p_d^{n_d}} \right|_{\mathbf{s}=\mathbf{p}}, n = n_1 + ... + n_d. \tag{6}$$

Since the FGC covariance function is statistically homogeneous and isotropic, the above condition simply requires the existence of the isotropic derivative of order $2n$ at zero pair separation distance, i.e. the existence of the following quantity

$$G^{(2n)}_{\mathrm{x};\lambda}(0) = (-1)^n \left[\frac{d^{2n} G_{\mathrm{x};\lambda}(\mathbf{r})}{dr^{2n}}\right]\Bigg|_{\mathbf{r}=\mathbf{0}} \tag{7}$$

Equation (7) is equivalent to the existence of the corresponding integral of the covariance spectral density

$$\left[\frac{d^{2n} G_{\mathrm{x};\lambda}(\mathbf{r})}{dr^{2n}}\right]\Bigg|_{\mathbf{r}=\mathbf{0}} = \eta_0\, \xi^d\, S_d \int_0^\infty dk\, \frac{\left|\tilde{Q}_\lambda(k)\right|^2 k^{d+2n-1}}{1+\eta_1\,(k\,\xi)^2 + (k\,\xi)^4} \tag{8}$$

where $S_d = \int d\hat{\mathbf{k}} = 2\pi^{d/2}/\Gamma(d/2)$ denotes the surface of the unit sphere in d dimensions. Note that if $\left|\tilde{Q}_\lambda(k)\right|^2 = 1$, i.e. in the absence of smoothing, the above integral does not exist unless $d+2n < 4$, which can be attained only for $d = 1$ and $n = 1$. If the smoothing kernel has a sharp cutoff k_c (band-limited spectrum), the $2n$-th order derivative is expressed in terms of the following integral

$$\frac{d^{2n} G_{\mathrm{x};\lambda}(\mathbf{r})}{dr^{2n}}\Bigg|_{\mathbf{r}=\mathbf{0}} = \eta_0\, \xi^{-2n}\, S_d \int_0^{k_c\xi} d\kappa\, \frac{\kappa^{d+2n-1}}{1+\eta_1\,\kappa^2+\kappa^4}. \tag{9}$$

The integral in 9 exists for all d and n. However, if the correlation length ξ exceeds significantly the resolution scale, i.e. $\xi >> \lambda$ and $k_c\xi >> 1$, for $\kappa >> 1$ the integrand behaves as κ^{d+2n-5}. Then, it follows $G^{(2n)}_{\mathrm{x};\lambda}(0) =$ regular + $\alpha_d\, \xi^{-2n} (k_c\xi)^{d+2n-4}$, where 'regular' represents the bounded contribution of the integral, while for fixed ξ the remaining term increases fast with $k_c\xi$. The constant α_d depends on the dimensionality of space. Hence, for $d \geq 2$ the singular term in $G^{(2n)}_{\mathrm{x};\lambda}(0)$ leads to large values of the covariance derivatives for $n \geq 1$. In [10] we focused on the case $k_c\xi >> 1$, which leads to 'rough' Spartan fields. Based on the above, the Gaussian FGC-SSRF can, at least in principle, interpolate between very smooth Gaussian random fields (e.g., Gaussian covariance function) and non-differentiable ones (e.g., exponential, spherical covariance functions). The 'degree' of smoothness depends on the value of the combined parameter $k_c\xi$. Hence, the FGC-SSRF in effect has four parameters, η_0, η_1, k_c, ξ , and the value of $k_c\xi$, which controls the smoothness of the model. This property of smoothness control is also shared by random fields with Matérn class covariance functions [14].

5 Non-Gaussian Probability Densities

An issue of significant practical importance is the ability of geostatistical models to capture fluctuations with non-Gaussian distributions. Such distributions can be developed in the Spartan-Gibbs framework by adding suitable (higher than second order) interaction terms in the energy functional. An example is the energy functional of the Landau model e.g. [10], which includes non-Gaussian terms and exhibits a transition between exponential and power-law spatial dependence of the covariance function. Geostatistical probability density models provide sufficient flexibility for fitting various types of non-Gaussian data. The approaches typically used in geostatistics for modeling asymmetric distributions with higher-than-normal weight in their tails employ the logarithmic and the Box-Cox transforms. In the former approach, the initial distribution is assumed to be approximately lognormal. The logarithmic mean $m_Y(\mathbf{s}) = E\left[\log X_\lambda(\mathbf{s})\right]$ is first estimated. Then, the fluctuations $y_\lambda(\mathbf{s}) = \log\left[X_\lambda(\mathbf{s})\right] - m_Y(\mathbf{s})$ follow the Gaussian distribution, and they can be modeled by means of the FGC-SSRF normalized energy density $h_{\mathrm{fgc}}\left[y_\lambda(\mathbf{s}); \eta_1, \xi\right]$. If the logarithm of the random field deviates from the Gaussian distribution, it is possible to modify the energy functional by adding a non-Gaussian term as follows

$$H_{\mathrm{ng}}\left[y_\lambda(\mathbf{s}); \eta_0, \eta_1, \xi, \mathbf{q}\right] = H_{\mathrm{fgc}}\left[y_\lambda(\mathbf{s}); \eta_0, \eta_1, \xi\right] + \delta H\left[y_\lambda(\mathbf{s}); \mathbf{q}\right], \qquad (10)$$

where δH is the non-Gaussian term that involves a parameter vector $\mathbf{q}$. For simplicity, below we are going to express (10) as $H = H_G + \delta H$, where H is the entire energy functional, and $H_G = H_{fgc}$ is the Gaussian FGC contribution. Now one has to determine the entire set of model parameters $\eta_0, \eta_1, \xi, \mathbf{q}$ (and possibly k_c) simultaneously from the sample. The deviation of the distribution from the Gaussian dependence is captured by means of additional constraints, e.g. based on the local terms $S_3(\mathbf{s}) = y_\lambda^3(\mathbf{s})$ and $S_4(\mathbf{s}) = y_\lambda^4(\mathbf{s})$. The corresponding distance functional then becomes

$$\begin{aligned} \Phi_s\left[X_\lambda(\mathbf{s})\right] = & \left|1 - \sqrt{\frac{\overline{S_1}}{\overline{S_0}} \frac{E[S_0]}{E[S_1]}}\right|^2 + \left|1 - \sqrt{\frac{\overline{S_2}}{\overline{S_0}} \frac{E[S_0]}{E[S_2]}}\right|^2 + \\ & \left|1 - \sqrt{\frac{\overline{S_3}}{\overline{S_0}^{3/2}} \frac{E[S_0]^{3/2}}{E[S_3]}}\right|^2 + \left|1 - \sqrt{\frac{\overline{S_4}}{\overline{S_0}^2} \frac{E[S_0]^2}{E[S_3]}}\right|^2 \end{aligned} \qquad (11)$$

The ratio $\overline{S_3} \big/ \overline{S_0}^{3/2}$ represents the sample skewness coefficient, while $\overline{S_4} \big/ \overline{S_0}^2$ the sample kurtosis coefficient. In the case of the Gaussian FGC-SSRF model, the stochastic moments $E\left[S_m\right]$, $m = 0, 1, 2$ (which are used in determining the model parameters) are expressed exactly in terms of the two-point covariance function. The covariance spectral density also follows directly from the energy functional. Such explicit expressions are not available for non-Gaussian energy functionals. The moments must be calculated either by numerical integration (e.g., Monte Carlo methods) for each set of parameters visited by the optimization method or by approximate, explicit methods that have been developed in the framework of many-body theories, e.g. [5, 8, 9, 13].

In statistical physics, e.g. [4, 5, 6] there is a long literature on approximate but explicit methods (variational approximations, Feynman diagrams, renormalization group, replicas) that address calculations with non-Gaussian probability densities. Preliminary efforts to apply these methods in geostatistical research [9, 12], and references therein] should be followed by further research on closed-form expressions for non-Gaussian Spartan densities and the accuracy of such approximations in various areas of the parameter space. In the variational approach [2, 5, 8], the non-Gaussian probability density is expanded around an 'optimal' Gaussian. The variational Gaussian can then be used as the zero-point approximation for low-order or diagrammatic perturbation expansions of the moments [9, 13]. Below, we outline the application of the variational method [2, 4, pp. 198–200, 5, pp. 71–77].

5.1 The Variational Method

We present the formalism of the variational method assuming that the SSRF is defined in a discretized space (e.g. on a lattice). The fluctuation random field and its states are denoted by the vector $\mathbf{y}$. The characteristic function $Z[\mathbf{J}]$ corresponding to the energy functional H is defined as

$$Z[\mathbf{J}] = \mathrm{Tr}\,\left[\exp(-H + \mathbf{J}\cdot\mathbf{y})\right]. \tag{12}$$

The symbol 'Tr' denotes the trace over all the field variables in H. For a lattice field the trace is obtained by integrating over the fluctuations at every point of the lattice. The cumulant generating functional (CGF) is defined by

$$F[\mathbf{J}] = -\log Z[\mathbf{J}]. \tag{13}$$

The cumulants of the distribution are obtained from the derivatives of the CGF with respect to $\mathbf{J}$. For example, the mean is given by

$$E\left[y(\mathbf{s}_i)\right] = -\left.\frac{\partial F[\mathbf{J}]}{\partial J_i}\right|_{\mathbf{J}=0}, \tag{14}$$

and the covariance function by

$$G_{\mathrm{y};\lambda}(\mathbf{s}_1,\mathbf{s}_2) = \left.\frac{\partial^2 F[\mathbf{J}]}{\partial J_1 \partial J_2}\right|_{\mathbf{J}=0}. \tag{15}$$

Higher-order cumulants are given by higher order derivatives of the CGF. The CGF of the Gaussian part $H_0 - \mathbf{J}\cdot\mathbf{y}$ is denoted as $F_0[\mathbf{J}]$. Let us now consider a variational Gaussian energy functional H_0, which is in general different than the Gaussian component H_G of H. The average of an operator A with respect to the pdf with energy H_0, is obtained by means of

$$\left\langle A\right\rangle_0 = \frac{\mathrm{Tr}\,A\,e^{-H_0}}{\mathrm{Tr}\,e^{-H_0}}. \tag{16}$$

The following inequality [5] is valid for all H_0

$$F[\mathbf{J}] \leq F_0[\mathbf{J}] + \left\langle H - H_0\right\rangle_0. \tag{17}$$

The optimal $\hat{H}_0$ that gives the best approximation of $F[\mathbf{J}]$, is obtained by minimizing the variational bound $F_0 + \langle H - H_0 \rangle_0$ with respect to the parameters of H_0. The optimal Gaussian pdf has energy $\hat{H}_0$ and provides approximate estimates of the non-Gaussian covariance function.

It is possible to improve on the variational approximation by expressing the energy functional H as follows

$$H = \hat{H}_0 + (H - \hat{H}_0) = \hat{H}_0 + (H_{\mathrm{G}} - \hat{H}_0 + \delta H), \tag{18}$$

and treating the component $H_{\mathrm{pert}} = H_{\mathrm{G}} - \hat{H}_0 + \delta H$ of the energy functional as a perturbation around the optimal Gaussian $\hat{H}_0$. Corrections of the stochastic moments can then be obtained either by means of simple (low-order) perturbation expansions, or by means of diagrammatic perturbation methods. However, there is no a priori guarantee that such corrections will lead to more accurate estimates, and such approximation must be investigated for each energy functional.

5.2 Example of Variational Calculation

Here we present a simple example for a univariate non-Gaussian pdf, which illustrates the application of the variational method. Consider the non-Gaussian energy functional

$$H(y) = a^2\, y^2 + \beta^4\, y^4, \tag{19}$$

where y is a fluctuation with variance $E[y^2]$, and the average is over the pdf $p(y) = Z^{-1} \exp(-H)$. The following Gaussian variational expression is used as an approximation of the non-Gaussian pdf

$$p_0(y) = \left(\sqrt{2\,\pi}\sigma\right)^{-1} \exp\left(-y^2/2\sigma^2\right). \tag{20}$$

Hence, the variational energy functional is $H_0 = y^2/2\sigma^2$ and σ is the variational parameter. It follows that $F_0 = -\log(\sqrt{2\,\pi}\sigma)$ and $\langle H - H_0 \rangle_0 = a^2\,\sigma^2 + 3\,\beta^4\,\sigma^4 - 1/2$. The variational bound given by (17) is a convex upward function of σ, as shown in Fig. 1. The bound is minimized for the following value of σ

$$\hat{\sigma} = \frac{\alpha}{6\beta^2} \left\{ 3\left[\sqrt{1 + 12\,\rho^4} - 1\right] \right\}^{1/2} = \frac{\alpha^{-1}}{6\rho^2} \left\{ 3\left[\sqrt{1 + 12\,\rho^4} - 1\right] \right\}^{1/2}. \tag{21}$$

In the above, $\rho = \frac{\beta}{\alpha}$ is the dimensionless ratio of the quartic over the quadratic pdf parameters that measures the deviation of the energy functional from the Gaussian form. The value of $\hat{\sigma}^2$ is the variational estimate of the variance. The exact variance, calculated by numerical integration, and the variational approximation for various values of the dimensionless coefficient ratio $\rho = \beta/\alpha$

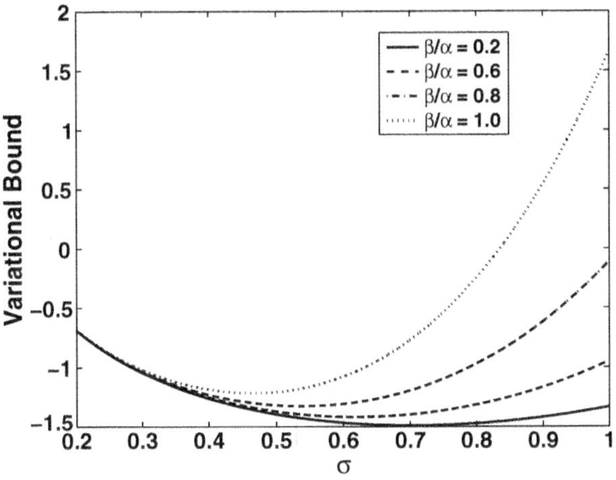

Fig. 1. Plots of the variational bound as a function of σ for four different values of the ratio β/α

are plotted in Fig. 2, which shows that the variational estimate is an excellent approximation of the exact result even for large values of the ratio ρ. Estimates based on first-order and cumulant perturbation expansions around the optimal Gaussian (these will be presented in detail elsewhere) are also shown in Fig. 2. The additional corrections do not significantly alter the outcome of the variational approximation for the variance, since all three plots almost coincide. However, such corrections will be necessary for calculating higher moments of non-Gaussian distributions. For example, the kurtosis of the

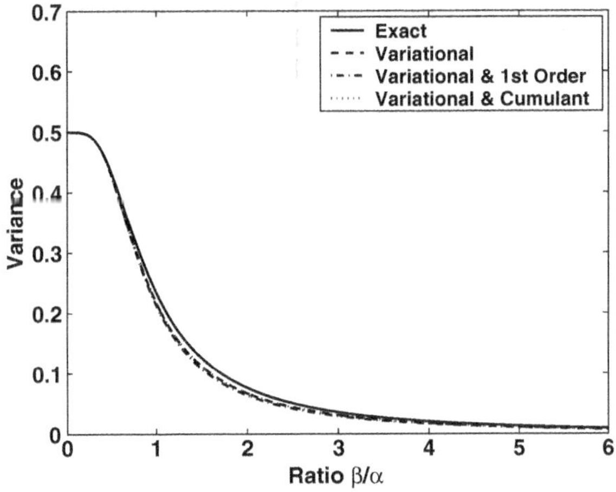

Fig. 2. Plots of the exact variance (numerical) and approximate estimates based on the variational approach as well as combinations of variational and perturbation methods (first order and cumulant expansion)

variational Gaussian is equal to 3, and thus it is not an accurate approximation of the kurtosis of the non-Gaussian distribution except for very small values of β.

6 Discussion

Spartan random fields provide an alternative to classical geostatistics for modeling the local variability of spatial processes. Spartan models are computationally efficient for large samples. In addition, they allow quantifying the variability of spatially sparse data sets, since the model parameters can be determined from a small number of measurements, in contrast with models based on variograms. The SSRFs also include a resolution scale that controls the smoothness of the field.

For SSRFs the structure of the energy functional, which may involve only short-range interactions, also determines the spatial dependence at large distances. In principle, the impact of this property on geostatistical modeling is mixed: On one hand, it does not allow estimating long-range dependence directly from the data. On the other hand, the estimation of the variogram at large distance often suffers from significant uncertainty due to insufficient number of pairs. Hence, the ability of SSRFs to model the long-range behavior of spatial processes needs to be investigated. It should also be mentioned that it is possible to modify the energy functional of the SSRFs by adding explicit long-range interactions.

Non-Gaussian distributions can be handled by means of the standard logarithmic transform. It is also possible to define interactions in the energy functional that lead to specific non-Gaussian probability densities. The complexity of the inference problem in this case increases compared to the Gaussian case. Certain methods that may be helpful for calculations with non-Gaussian densities were suggested in this paper, and the variational method was presented in more detail with the help of a specific univariate example.

Certain other methodological and numerical issues of SSRFs require further investigation. The methodological issues include estimation at unsampled points, Monte Carlo simulation, application to real data sets, formulation of estimation uncertainty, stability of model parameters to uncorrelated noise, modelling of spatial processes with multiple scales of variability and anisotropic structures. Estimation has been briefly discussed in [10], and unconditional simulation in [11]. Numerical issues involve efficient algorithms for optimization (model inference process), simulation, and the processing of spatial information in problems with irregular supports.

References

1. Adler RJ (1981) The Geometry of Random Fields. Wiley, New York
2. Barthelemy M, Orland H, Zerah G (1995) Propagation in random media: calculation of the effective dispersive permittivity by use of the replica method. Phys Rev E 52(1):1123–1127
3. Bochner S (1959) Lectures on Fourier Integrals. Princeton University Press, Princeton, NJ
4. Chaikin PM, Lubensky TC (1995) Principles of Condensed Matter Physics. Cambridge University Press, UK
5. Feynman RP (1982) Statistical Mechanics. Benjamin and Cummings, Reading, MA
6. Goldenfeld N (1993) Lectures on Phase Transitions and the Renormalization Group. Addison-Wesley, New York
7. Goovaerts P (1997) Geostatistics for Natural Resources Evaluation. Oxford University Press, NY
8. Hristopulos DT, Christakos G (1997) A variational calculation of the effective fluid permeability of heterogeneous media. Phys Rev E 55(6): 7288–7298
9. Hristopulos DT, Christakos G (2001) Practical calculation of non-Gaussian multivariate moments in spatiotemporal Bayesian maximum entropy analysis. Math Geol 33(5): 543–568
10. Hristopulos DT (2003) Spartan Gibbs random field models for geostatistical applications. SIAM J Sci Comp 24: 2125–2162
11. Hristopulos DT (2003) Simulations of spartan random fields. In: Simos TE (ed) Proceedings of the international conference of computational methods in sciences and engineering 2003: 242–247. World Scientific, London, UK
12. Hristopulos DT (2003) Renormalization group methods in subsurface hydrology: Overview and applications in hydraulic conductivity upscaling. Adv Water Resour 26(12): 1279–1308
13. Meurice Y (2002) Simple method to make asymptotic series of Feynman diagrams converge. Phys Rev Lett 88(14):1601–1604
14. Pilz J (2003) Bayesian spatial prediction using the Matern class of covariance models. In: Dubois G, Malczewski J, De Cort M (eds) Mapping radioactivity in the environment: Spatial interpolation comparison 1997: 238–252. Office for Official Publications of the European Communities, Luxembourg
15. Yaglom M (1987) Correlation Theory of Stationary and Related Random Functions I. Springer, New York

Bayesian Trans-Gaussian Kriging with Log-Log Transformed Skew Data

Gunter Spöck, Hannes Kazianka, and Jürgen Pilz

Department of Statistics, University of Klagenfurt, Klagenfurt, Austria
gunter.spoeck@uni-klu.ac.at
hannes.kazianka@uni-klu.ac.at
juergen.pilz@uni-klu.ac.at

1 Introduction

Besides the assumption of stationarity it is conventional among geostatistical practitioners to make the assumption of Gaussianity when applying the linear kriging methodology. The standard procedure in geostatistics is to

- estimate an empirical variogram or covariance function,
- fit a theoretical variogram or covariance model to the empirical estimate by means of least squares,
- apply linear kriging to the data by plugging in the estimated positive semi-definite theoretical covariance model,
- report uncertainties of the predictions by means of the plug-in kriging variance and Gaussian-based confidence intervals.

The above approach has certain deficiencies:

- The empirical variogram or covariance estimates at the different lags are highly correlated and, therefore, can be very misleading.
- Once the empirical variogram is badly estimated, the same is also true for the fitted theoretical model.
- Plugging the estimated theoretical covariance model in the kriging predictor and neglecting the uncertainty of the covariance estimate has the following consequences:
  - the plug-in kriging predictor is neither a linear nor best predictor.
  - the kriging variance that is based on a plug-in estimate is not the true variance of this highly non-linear plug-in kriging predictor and actually underestimates the true unknown variance [5].
- In hardly any application the assumption of Gaussianity is fulfilled. Most environmental processes like ore grades, soil and air measurements show highly right-skewed marginal distributions.

All these negative facts were the motivation for us to look for more advanced kriging methodologies that relax the Gaussian assumption and the disadvantage of not taking into account the uncertainty of the covariance function.

One of the first papers that addressed the above deficiencies and influenced our work was De Oliveira [15]. He developed a Bayesian trans-Gaussian kriging where uncertainties could be specified on the trend, the covariance function and the parameter of the transformation function. The transformation he used to make a skew random field Gaussian was the Box-Cox transformation. Motivated by the conjugacy of the normal-inverse-gamma family to the normal sampling distribution he exactly used such kind of prior for the sill, the range and the trend parameters of his model. His approach distinguishes from our approach by the fact that his prior is informative. Berger et al. [3] were the first to investigate also non-informative reference priors for correlated stochastic processes. The disadvantage of their approach is that the assumed random field must be Gaussian. Some time later the paper [16] appeared where also a non-informative prior for the transformed Gaussian model was proposed, but with fixed range parameter while the sill and the nugget are variable.

Our approach seems to be the first completely non-informative approach to trans-Gaussian Bayesian kriging. We avoid the specification of a prior distribution by means of a parametric bootstrap, where the sampling distribution of maximum likelihood estimates is taken as the posterior for unknown parameters.

2 Bayesian Trans-Gaussian Kriging Model

Throughout the paper we apply the convention that the underlying random field is given by $\{Z(x) : x \in \mathcal{D}\}$ whereas the transformed Gaussian random field is denoted by $Y(x) = \mathrm{g}\left(Z(x)\right)$. One transformation that is also conventional in regression analysis is the Box-Cox transformation defined for positive data z and given by

$$g_\lambda(z) = \begin{cases} \frac{z^\lambda - 1}{\lambda} & : \lambda \neq 0 \\ \log(z) & : \lambda = 0 \end{cases},$$

This transformation was used by De Oliveira [15] and Pilz and Spoeck [19] in their approach to trans-Gaussian Bayesian kriging. Log-normal kriging forms a special case, where the log-transform of the underlying random field $\{Z(x) : x \in \mathcal{D}\}$ is assumed to be Gaussian: $Y(x) = \log Z(x) \sim N(m(x), \sigma^2)$ where $\sigma^2 = \mathrm{var}\{Y(x)\}$ is the (constant) variance of the log-transformed random field. The Box-Cox transformation is able to model moderately skewed marginal distribution.

The innovation of this paper is that we apply a new kind of transformation that we call the log-log transformation and the fact that the work of Pilz et al.

[19] is extended to allow also for uncertainty in geometric anisotropy parameters. See Appendix for details. The log-log transformation is only defined for positive data z and can be written as

$$g_\lambda(z) = \log(\log(z) + \lambda) \,. \tag{1}$$

The restriction $\log(z) > -\lambda$ for all z out of the support of the random process must be fulfilled. This restriction is not relevant in applications since we only have a finite set of observations and thus $\min_z \log(z) > -\infty$. The main advantage of the log-log transformation is that because of the double logarithm highly skewed data can be potentially transformed to a normal distribution.

For the trend and error model of the transformed Gaussian random field we assume the conventional geostatistical model:

$$\mathrm{E}\{Y(x)\} = \mu,$$

where μ is a constant trend. For the Gaussian error model we assume a covariance function C_{θ,σ^2} of the form

$$C_{\theta,\sigma^2}(x_1 - x_2) = \sigma^2 k_\theta(x_1 - x_2) \tag{2}$$

where $\sigma^2 = \mathrm{var}\{Y(x)\}$ denotes the variance (overall sill) of the random field and $k_\theta(\cdot)$ the correlation function (normalized covariance function); $\theta \in \Theta \subset \mathbb{R}^m$ stands for a parameter vector whose components describe the range and shape of the positive definite correlation function. Under these assumptions the probability density of the observed data takes the form

$$\begin{aligned} f(\mathbf{Z}; \mu, \theta, \sigma^2, \lambda) &= J_\lambda(\mathbf{Z}) * ((2\pi)^n \det(\sigma^2 \mathbf{K}_\theta))^{-\frac{1}{2}} \\ &\quad * \exp\{-\frac{1}{2}(g_\lambda(\mathbf{Z}) - \mathbf{1}\mu)^T (\sigma^2 \mathbf{K}_\theta)^{-1} (g_\lambda(\mathbf{Z}) - \mathbf{1}\mu)\}, \end{aligned}$$

where $J_\lambda(\mathbf{Z})$ is the determinant of the Jacobian of the specific transformation used and $\sigma^2 \mathbf{K}_\theta$ is the covariance matrix of

$$\mathbf{Y} = g_\lambda(\mathbf{Z}) = (g_\lambda(Z(x_1)), g_\lambda(Z(x_2)), \ldots, g_\lambda(Z(x_n)))^T.$$

We consider the transformation parameter λ to be unknown and to be estimated. Christensen et al. [6] point out that the interpretation of the μ-parameter changes with the value of λ, and the same applies to the covariance parameters σ^2 and θ. Estimation of the parameter λ can be performed by a profile likelihood approach. For fixed values of λ and θ, the maximum likelihood estimates for μ and σ^2 are given by

$$\hat{\mu}^{OK}_{\lambda,\theta} = (\mathbf{1}^T {\mathbf{K}_\theta}^{-1} \mathbf{1})^{-1} \mathbf{1}^T {\mathbf{K}_\theta}^{-1} g_\lambda(\mathbf{Z}), \tag{3}$$

and

$$\hat{\sigma}^2_{\lambda,\theta} = \frac{1}{n-1}(g_\lambda(\mathbf{Z}) - \mathbf{1}\hat{\mu}^{OK}_{\lambda,\theta})^T {\mathbf{K}_\theta}^{-1} (g_\lambda(\mathbf{Z}) - \mathbf{1}\hat{\mu}^{OK}_{\lambda,\theta}),$$

respectively, where $\mathbf{K}_\theta$ is the correlation matrix. The estimates for λ and θ have no closed form expression and have to be found numerically by plugging $\hat{\mu}^{OK}_{\lambda,\theta}$ and $\hat{\sigma}^2_{\lambda,\theta}$ in the above likelihood function for numerical maximization.

The prediction problem can be faced by the help of the conditional mean. The conditional mean – or ordinary kriging predictor – $E\{Z(x_0)|\mathbf{Z}\}$, is optimal with respect to the squared error loss. Moreover, it may be shown that the conditional distribution, $[Y(x_0)|\mathbf{Y}]$, is normal with mean and variance equal to this ordinary kriging predictor and kriging variance on the transformed scale.

Since we want to apply the Bayesian paradigm we have to introduce appropriate prior distributions for the mean parameter μ and the covariance parameters θ and σ^2. The first who made use of the Bayesian approach in kriging were Kitanidis [13], Omre [17] and Omre and Halvorsen [18]. Contrary to us, they assumed a Gaussian random function model for $Z(\cdot)$ with fixed, exactly known covariance function and only incorporated probabilistic prior information for the trend function. The Bayesian ordinary kriging predictor may be written as

$$\hat{Z}^{\theta,\sigma^2}_{BK}(x_0) = \hat{\mu}^{\theta,\sigma^2}_{BK} + \mathbf{c}_\theta^T \mathbf{K}_\theta^{-1}(\mathbf{Z} - \mathbf{1}\hat{\mu}^{\theta,\sigma^2}_{BK}), \tag{4}$$

$$\hat{\mu}^{\theta,\sigma^2}_{BK} = (\mathbf{1}^T\mathbf{K}_\theta^{-1}\mathbf{1} + \Phi^{-1})^{-1}(\mathbf{1}^T\mathbf{K}_\theta^{-1}\mathbf{Z} + \Phi^{-1}\mu_0), \tag{5}$$

where $\mathrm{E}(\mu) = \mu_0$ is the fixed a-priori mean and $\mathrm{var}(\mu) = \sigma^2\Phi$ is the fixed a-priori variance for μ. The vector $\mathbf{c}_\theta$ contains the correlations between the point to be predicted and the observations at the n locations. It can be shown that the total mean-squared error (TMSEP) of this predictor

$$\mathrm{E}\{Z^{\theta,\sigma^2}_{BK}(x_0) - Z(x_0)\}^2 =$$
$$\sigma^2\left(1 - \mathbf{c}_\theta^T\mathbf{K}_\theta^{-1}\mathbf{c}_\theta + ||1 - \mathbf{1}^T\mathbf{K}_\theta^{-1}\mathbf{c}_\theta||^2_{(\mathbf{1}^T\mathbf{K}_\theta^{-1}\mathbf{1})^{-1}}\right),$$

where $||\mathbf{a}||^2_{\mathbf{A}}$ is a short-hand for the quadratic form $\mathbf{a}^T\mathbf{A}\mathbf{a}$, is always smaller than the mean-squared-error of prediction (MSEP) of the ordinary kriging predictor. Thus, by accepting a small bias in the Bayes kriging predictor and using prior knowledge $\mathrm{E}(\mu) = \mu_0$ and $\mathrm{var}(\mu) = \sigma^2\Phi$ one gets better predictions than with ordinary kriging. We refer to Spöck [20], where these results are investigated in more detail.

An obvious advantage of the Bayesian approach, besides its ability to deal with the uncertainty of the model parameters, is the compensation for the lack of information in case of only few measurements. This has been demonstrated impressively by Omre [17], Omre and Halvorsen [18] and Abrahamsen [1].

Bayesian linear kriging is not fully Bayesian, since it makes no a-priori distributional assumptions on the parameters of the covariance function. The first to take also account of the uncertainty with respect to these parameters, using a Bayesian setup, were Kitanidis [13] and Handcock and Stein [11]. A prior different from the one of Handcock and Stein was used by Gaudard et al.

[10]. Further references to Bayesian spatial prediction approaches are Le and Zidek [14], Handcock and Wallis [12], Cui et al. [7], Ecker and Gelfand [9] and Banerjee et al. [2].

We now come to discuss the Bayesian approach to trans-Gaussian kriging. Whereas, in conventional trans-Gaussian kriging [6] the uncertainty of the transformation to Gaussianity and the uncertainty of the covariance function was not considered, De Oliveira et al. [15] have proposed a Bayesian trans-Gaussian kriging method which takes full account of these uncertainties. A prior $p(\mu, \theta, \sigma^2, \lambda)$ is specified for all unknown parameters. "But the choice of the prior distribution requires some care, because the interpretation of μ, σ^2 and θ depends on the realized value of λ. Each transformation (i.e. each λ) will change the location and scale of the transformed data, as well as the correlation structure, so assuming them to be independent a priori of λ would give nonsensical results", (citation from De Oliveira et al. [15]). Defining $\tau = \frac{1}{\sigma^2}$, their full prior specification is based on a proposal of Box and Cox [4] and is given by the improper density

$$p(\mu, \theta, \tau, \lambda) = \frac{p(\theta)p(\lambda)}{\tau J_\lambda^{1/n}(\mathbf{Z})}.$$

Observe, this prior is dependent on the data $Z(x_i)$, $i = 1, 2, \ldots, n$. De Oliveira et al. [15] used their method for the spatial prediction of weekly rainfall amounts. "It performed adequately and slightly better than several kriging variants, especially regarding the empirical probability of coverage of the nominal 95% prediction intervals", (citation from De Oliveira et al. [15]).

3 The Bootstrap Approach

Our empirical Bayesian approach to trans-Gaussian kriging has delicate differences to the approach taken by De Oliveira et al. [15]. They have to specify priors $p(\theta)$ and $p(\lambda)$ for covariance and transformation parameters. Up to date there exists only very preliminary work on non-informative priors for these parameters (Berger et al. [3]). So, nobody actually knows what the information content of informative priors is. Our approach is a kind of empirical Bayes and non-informative step to reasoning using some Bayesian paradigm. For the mean parameter μ we use a normal prior $\mu \sim \mathcal{N}\left(\mu_0, \sigma^2 \Phi\right)$ but instead of specifying prior distributions for θ, σ^2 and λ we implicitly use only information from the data by means of parametrically bootstrapping ML-estimators of these parameters. We then treat this bootstrap distribution as posterior and apply the Bayesian predictive approach.

Generally, the parametric bootstrap of these ML-estimators works as follows:

1. From the original data $Z(x_1), \ldots, Z(x_n)$ the desired covariance and transformation parameters are estimated by means of ML to get $\hat{\theta}_0, \hat{\sigma}_0^2$ and $\hat{\lambda}_0$.

2. The estimated parameters are subsequently used for simulating transformed random fields as follows: With the estimated covariance parameters a Gaussian random field with mean equal to the initial generalized least squares estimate $\hat{\mu}_{BK}^{\hat{\theta}_0,\hat{\sigma}_0^2}$ is generated on the locations $x_1, \ldots, x_n$ and the estimated transformation parameter, $\hat{\lambda}_0$, is then used to transform the random field to a trans-Gaussian one. Arbitrary strictly monotone transformations g_λ to Gaussianity and back-transformations may be used. Special cases would be the Box-Cox or the log-log transformation.
3. The second step is repeated for a large number N of simulations and the covariance and transformation parameters are re-estimated from the simulated data by means of ML.
4. The result of this parametric bootstrap is a set of ML estimates $\left(\hat{\sigma}_i^2, \hat{\theta}_i, \hat{\lambda}_i\right)_{i=1,\ldots,N}$.

The bootstrap distribution reflects the uncertainty of the covariance function and the correct transformation to Gaussianity. Treating this bootstrap distribution as posterior for the unknown parameters in the Bayesian predictive approach thus takes account of all mentioned uncertainties. Since we have simulations from our posterior distribution we can proceed by means of a Monte-Carlo approximation to the predictive distribution at unknown locations z_0:

$$p(z_0|\mathbf{Z}) \simeq \frac{1}{N} \sum_{i=1}^{N} p(g_{\hat{\lambda}_i}(z_0)|\hat{\lambda}_i, \hat{\theta}_i, \hat{\sigma}_i^2, \mathbf{Z}) * J_{\hat{\lambda}_i}(z_0) \tag{6}$$

Here $p(g_{\hat{\lambda}_i}(z_0)|\hat{\lambda}_i, \hat{\theta}_i, \hat{\sigma}_i^2, \mathbf{Z})$ is the conditional predictive density,

$$Y(x_0)|\hat{\lambda}_i, \hat{\theta}_i, \hat{\sigma}_i^2, \mathbf{Z} \sim \mathcal{N}(\hat{Y}_{BK}^{\hat{\lambda}_i,\hat{\theta}_i,\hat{\sigma}_i^2}(x_0), TMSEP_{\hat{\lambda}_i,\hat{\theta}_i,\hat{\sigma}_i^2}),$$

where $\hat{Y}_{BK}^{\hat{\lambda}_i,\hat{\theta}_i,\hat{\sigma}_i^2}(x_0)$ is the Bayes kriging predictor applied to the transformed data $\mathbf{Y} = g_{\hat{\lambda}_i}(\mathbf{Z})$ for fixed $(\hat{\lambda}_i, \hat{\theta}_i, \hat{\sigma}_i^2)$, and $TMSEP_{\hat{\lambda}_i,\hat{\theta}_i,\hat{\sigma}_i^2}$ is the corresponding Bayes kriging variance. From this predictive distribution quantiles, the median, the mean and probabilities above certain thresholds can easily be calculated.

4 Application to the SIC2004 Joker Data Set

One of our aims is to have a methodology that is intrinsic Bayesian and can be applied also to highly skewed data sets that often occur in applications. In 2004 one such data set was investigated in detail during the spatial interpolation contest (SIC2004) [8]. Ten training data sets on radioactivity levels were given to the participants of the contest in a certain period of time to train their automatic interpolation algorithms. After training another data set, called "Joker", was given to the probands, which had one completely different property than the training data. "A small corner located SW of the monitored area was chosen and a dispersion process was modelled in order to obtain a few values on the order of 10 times more than the overall background levels reported for the first data set", according to [8]. The automatic interpolation routines applied to this data ranged from ordinary kriging, splines, support vector machines to neural networks. The performance of the different algorithms could later on be compared to the true values. Performance measures such as mean absolute error (MAE) and root mean squared error (RMSE) have been reported and published in [8]. Because we already know the true data when we have looked at the performance of our Bayesian trans-Gaussian kriging algorithm, the calculations we give here are outside the context of a competition. But our results are in performance comparable to the best real geostatistical algorithm from this contest. The winner was a neural network algorithm.

Figure 1(a) gives the data locations of the joker data set (blue circles) together with the locations where the prediction should take place (red stars). For an exploratory data analysis we refer to Dubois [8]. A histogram of the Joker data set is shown in Fig. 1(b). The histogram shows that the background level is quite symmetric, however, there exist also some very large values that can be interpreted as an accidental release of radioactivity.

The methodology we apply to this data set is Bayesian trans-Gaussian kriging with the log-log transformation. Since we are not sure whether the variogram model is linear or parabolic in the origin we have used a convex combination of a Gaussian and an exponential variogram model. The advantage of our method is that according to the data the bootstrap methodology then takes account also of this uncertainty of the variogram model in the origin. The convex combination parameter as well as Gaussian and exponential range parameters, the overall sill and transformation parameter are part of the bootstrap. The anisotropy is respected as well by including a transformation matrix for the coordinates in the maximum likelihood bootstrap. As already mentioned the main advantage of our approach is that all uncertainties are taken into account and no prior specification is necessary. Figure 2(a)–(d) show the bootstraped transformation parameters, covariance parameters and variogram functions from the posterior. Although estimation of geometric anisotropy was performed, it turned out that the boostrapped semivariogram realizations show no anisotropy. Because we calculate posterior predictive distributions (see Fig. 3) at all locations where prediction should take place by means of Monte Carlo averaging with the samples from the bootstrap, graphics like quantile maps, Fig. 4 (a)–(d) and Fig. 5(b), posterior mean map, Fig. 5(a), and maps of the probability above thresholds, Fig. 6, are available. To make our results comparable to the SIC2004 contest we calculated the MAE=16.19 and RMSE=77.64. In terms of MSE this would have been the second best result in the SIC2004 contest.

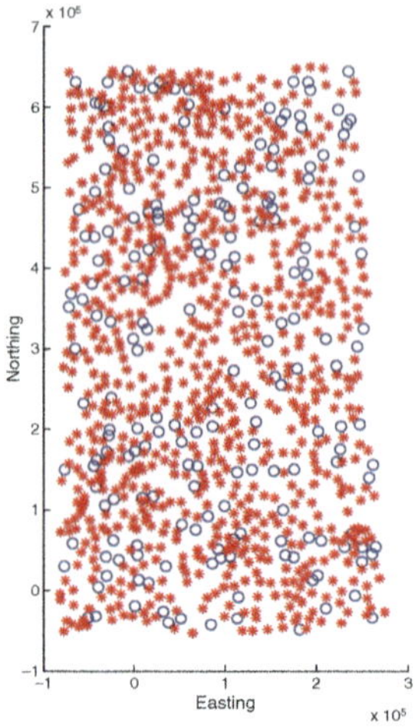

(a)

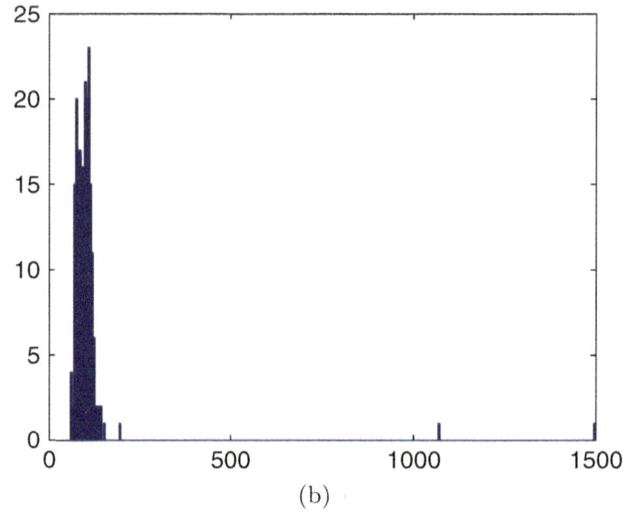

(b)

Fig. 1. (a) The data locations (*blue circles*) and the locations where prediction takes places (*red stars*). **(b)** The histogram of the Joker data set

Because in most practical applications true data at the locations where prediction is desired are not available, we have tested the performance of our algorithm also in the context of cross-validation. Results are shown in Fig. 7. Furthermore, we must mark that the Bayesian posterior mean is conditionally biased as was expected. Small values are predicted quite well but the extreme values from the dispersion process are underestimated. This has been expected since only two very large values are included in the training set. Validation results with the true data values are similar.

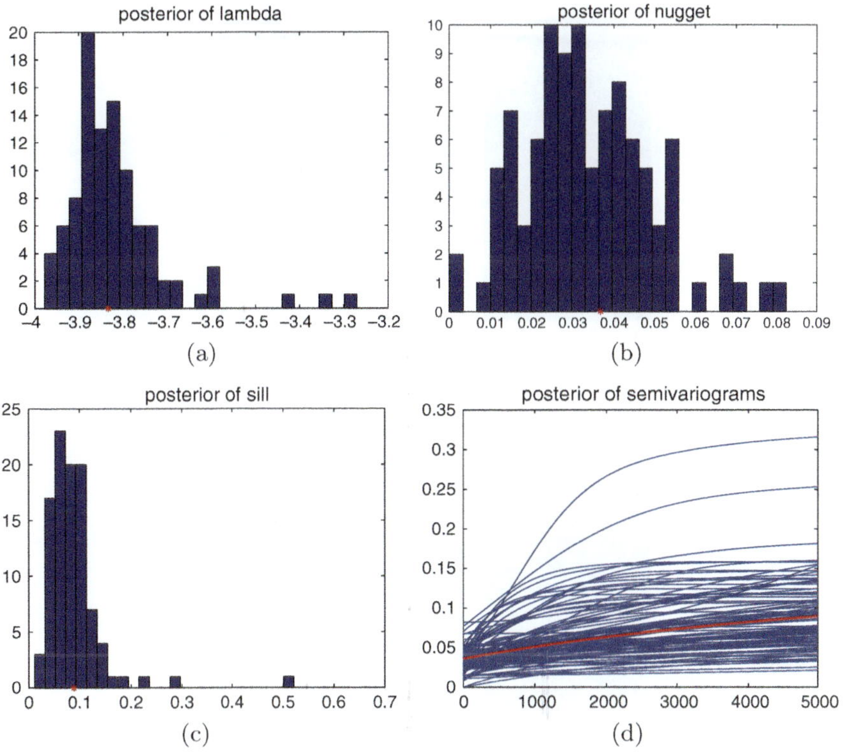

Fig. 2. Results of parameteric boostrap taken as posterior. **(a)** posterior distribution of transformation parameter, **(b)** posterior of nugget, **(c)** posterior of sill, **(d)** semivariogram functions. *Red stars* and *lines* indicate the initial ML estimates on which the parametric bootstrap is based

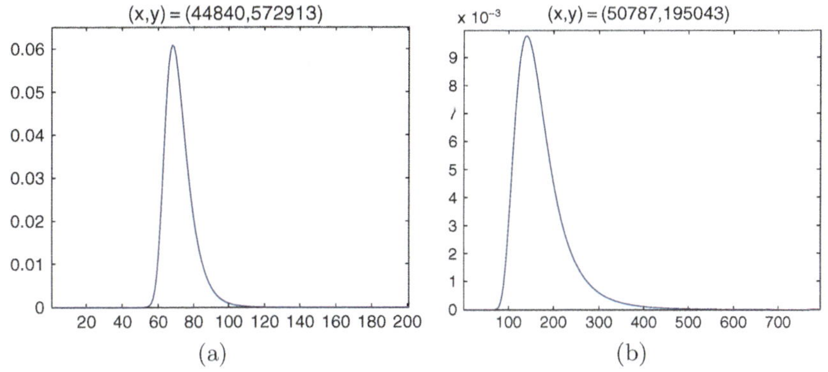

Fig. 3. **(a)** Posterior predictive distribution at a "normal" background location. **(b)** posterior predictive distribution at a "hotspot"

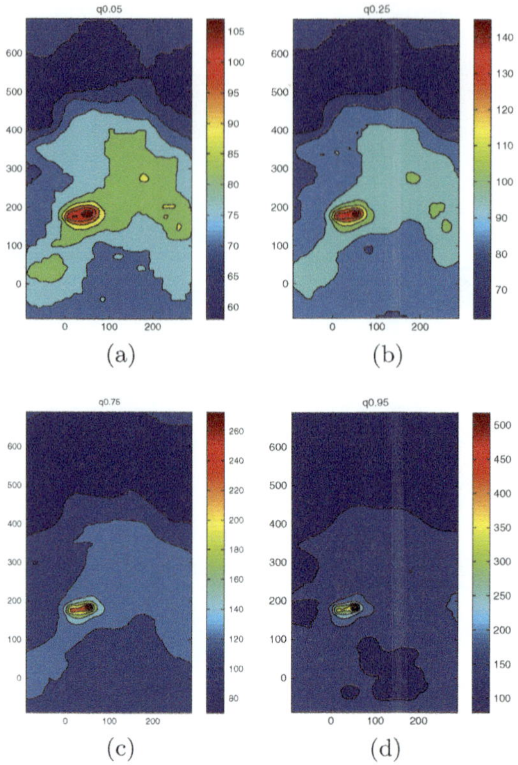

Fig. 4. Maps of the quantiles of the posterior predictive distribution. **(a)** 5% quantile, **(b)** 25% quantile, **(c)** 75% quantile, **(d)** 95% quantile

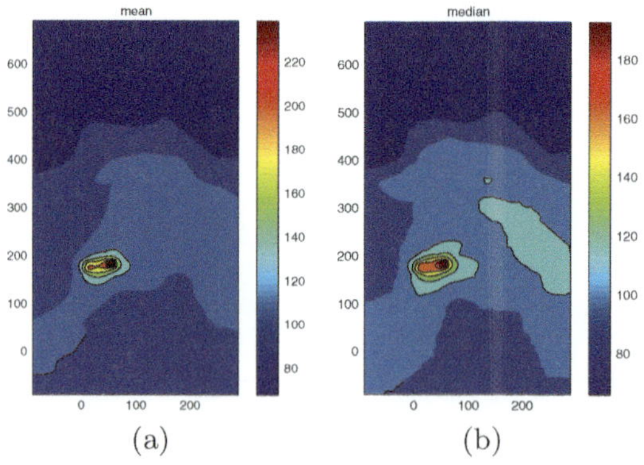

Fig. 5. **(a)** posterior predictive mean. **(b)** posterior predictive median

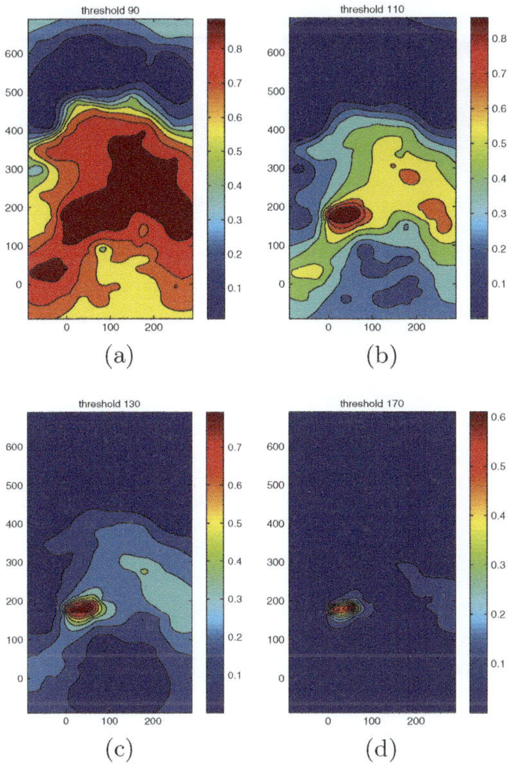

Fig. 6. Maps of the probabilities above certain thresholds. **(a)** threshold 90, **(b)** threshold 110, **(c)** threshold 130, **(d)** threshold 170

5 Conclusion

Skewed, non-Gaussian data are common in environmental and geostatistical applications. Conventional geostatistical methods do not take into account these facts. The present paper investigates a method to deal with this sort of data. Because conventional kriging methodology does not respect the fact that the used covariance function is an estimate and therefore always is uncertain to some degree, our aim was to propose an approach that takes the uncertainty of the covariance estimates into account. The suggested algorithms are Bayesian in spirit but avoid the tedious specification of prior distributions for unknown parameters for which up to date no, in the non-Gaussian context, practically useful work exists. By applying a bootstrapping procedure and interpreting the bootstrap distribution as posterior distribution we result in an empirical Bayes method. A data example taken from the SIC2004 contest demonstrates the usefulness of our methodology.

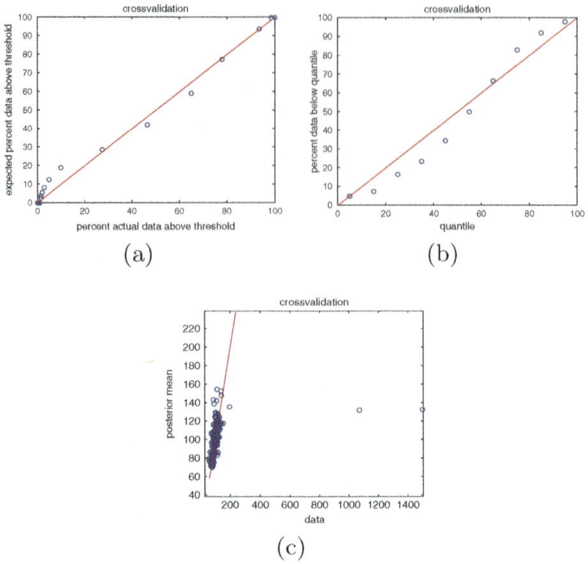

Fig. 7. Crossvalidation results. **(a)** percent of actual data vs. expected percent of data above the thresholds $10, 20, \ldots, 170, 1000, 1100, \ldots, 1500$. **(b)** posterior predictive quantiles vs. percent of data below quantiles. **(c)** data vs. posterior predictive mean

Appendix: Details on ML Covariance Estimation

For ML estimation of the parameters θ, σ^2 and λ we have to maximize the likelihood function given by

$$L(\mu, \lambda, \sigma^2, \theta; \mathbf{Z}) = p(g_\lambda(\mathbf{Z})|\mu, \sigma^2, \lambda, \theta) * J_\lambda(\mathbf{Z}),$$

where the Jacobian is given by

$$J_\lambda(\mathbf{Z}) = |\det(\frac{\partial}{\partial \lambda} g_\lambda(\mathbf{Z}))|.$$

and the density of the transformed data is

$$p(g_\lambda(\mathbf{Z})|\mu, \sigma^2, \lambda, \theta) = \mathcal{N}\{\mathbf{1}\mu, \sigma^2 \mathbf{K}_\theta; g_\lambda(\mathbf{Z})\}.$$

Because we have prior knowledge about β in the form $\mathrm{E}(\mu) = \mu_0, \mathrm{var}(\mu) = \sigma^2 \Phi$ it makes sense to use the profile-likelihood instead:

$$L(\lambda, \sigma^2, \theta; \mathbf{Z}) = p(g_\lambda(\mathbf{Z})|\hat{\mu}_{BK}^{\theta,\sigma^2}, \sigma^2, \lambda, \theta) * J_\lambda(\mathbf{Z}),$$

where

$$p(g_\lambda(\mathbf{Z})|\hat{\mu}_{BK}^{\theta,\sigma^2}, \sigma^2, \lambda, \theta) = \mathcal{N}\{\mathbf{1}\hat{\mu}_{BK}^{\theta,\sigma^2}, \sigma^2 \mathbf{K}_\theta; g_\lambda(\mathbf{Z})\}.$$

Numerical routines like Gauss-Newton but also Line-Search are sensitive to the starting values $(\lambda_0, \sigma_0^2, \theta_0)$ and to maximizing the likelihood for $(\hat{\lambda}, \hat{\sigma}^2, \hat{\theta})$ at once. We have therefore adopted a profile-likelihood approach and used the standard Matlab routines for maximization.

1. First estimate $\hat{\lambda}_0$ for λ: Least squares fitting of a Gaussian density to a kernel density estimate of the transformed data.
2. First estimate for the covariance function:
   (a) Transform the data with $\hat{\lambda}_0$.
   (b) Calculation of an anisotropic empirical variogram estimate from the transformed data.
   (c) Fitting of a theoretical anisotropic variogram model to the empirical variogram to get estimates $\hat{\sigma}_0^2, \hat{\theta}_0$ and linear anisotropy transformation matrix $\hat{A}_0$.
3. Repeat the following steps for $i = 0, 1, 2, \ldots$ until convergence of the estimates:
   (a) Fix $\sigma^2 = \hat{\sigma}_i^2, \theta = \hat{\theta}_i, A = \hat{A}_i$ and maximize the profile-loglikelihood $l(\lambda, \sigma^2, \theta, A; \mathbf{Z})$ for λ to get $\hat{\lambda}_{i+1}$.
   (b) Fix $\lambda = \hat{\lambda}_{i+1}, A = \hat{A}_i$ and maximize the profile-loglikelihood $l(\lambda, \sigma^2, \theta, A; \mathbf{Z})$ for σ^2, θ to get $\hat{\sigma}_{i+1}^2, \hat{\theta}_{i+1}$.
   (c) Fix $\lambda = \hat{\lambda}_{i+1}, \sigma^2 = \hat{\sigma}_{i+1}^2, \theta = \hat{\theta}_{i+1}$ and maximize the profile-loglikelihood $l(\lambda, \sigma^2, \theta, A; \mathbf{Z})$ for the anisotropy transformation matrix A to get $\hat{A}_{i+1}$.

References

1. Abrahamsen, P. (1992), "Bayesian kriging for seismic depth conversion of a multilayer reservoir", in Geostatistics Troia '92, A. Soares ed., Kluwer, Dordrecht, 358–398
2. Banjeree, S., Carlin, B. & Gelfand, A. (2004), "Hierarchical modelling and analysis for spatial data", Chapman and Hall/CRC, Boca Raton, Florida
3. Berger, J., De Oliveira, V. & Sanson, B. (2001), "Objective Bayesian analysis of spatially correlated data", Journal of the American Statistical Association, 96, 456, 1361–1374
4. Box, G. & Cox, D. (1964), "An analysis of transformations", Journal of the Royal Statistical Society, B, 26, 211–252
5. Christensen, R. (1991), "Linear models for multivariate time series and spatial data", Springer, Berlin
6. Christensen, O., Diggle, P. & Ribeiro, P. (2001), "Analysing positive-valued spatial data: The transformed Gaussian model", in GeoENV III – Geostatistics for Environmental Applications, eds. P. Monestiez, D. Allard and R. Froidevaux, Kluwer, Dordrecht, 287–298
7. Cui, H., Stein, A. & Myers, D. (1995), "Extension of sptial information, Bayesian kriging and updating of prior variogram parameters", Environmetrics, 6, 373–384
8. Dubois, G. & Galmarini, S. (2005), "Spatial interpolation comparison (SIC) 2004: Introduction to the exercise and overview of results", in Automatic Mapping Algorithms for Routine and Emergency Monitoring Data, ed. G. Dubois, Belgium
9. Ecker, M. & Gelfand, A. (1997), "Bayesian variogram modelling for an isotropic spatial process", Technical Report 97-01, Department of Statistics, University of Connecticut
10. Goudard, M., Karson, M., Linder, E. & Sinha, D. (1999), "Bayesian spatial prediction", Environmental and Ecological Statistics, 6, 147–171
11. Handcock, M. & Stein, M. (1993), "A Bayesian analysis of kriging", Technometrics, 35, 4, 403–410
12. Handcock, M. & Wallis, J. (1994), "An approach to statistical spatio-temporal modelling of meteorological fields", Journal of the American Statistical Association, 89, 368–378

13. Kitanidis, P. (1986), "Parameter uncertainty in estimation of spatial functions: Bayesian analysis", Water Resources Research, 22, 499–507
14. Le, N. & Zidek, J. (1992), "Interpolation with uncertain spatial covariance: A Bayesian alternative to kriging", Journal of Multivariate Analysis, 43, 351–374
15. De Oliveira, V., Kedem, B. & Short, D. (1997), "Bayesian prediction of transformed Gaussian random fields", Journal of the American Statistical Association, 92, 440, 1422–1433
16. De Oliveira, V. (2007), "Objective Bayesian analysis of spatial data with measurement error", Canadian Journal of Statistics, 35, 283–301
17. Omre, H. (1987), "Bayesian kriging-merging observations and qualified guess in kriging", Mathematical Geology, 19, 25–39
18. Omre, H. & Halvorsen, K. (1989), "The Bayesian bridge between simple and universal kriging", Mathematical Geology, 21, 7, 767–786
19. Pilz, J. & Spoeck, G. (2007), "Why do we need and how should we implement Bayesian kriging methods", Stochastic Environmental Research and Risk Assessment, DOI: 10.1007/s00477-007-0165-7, Springer
20. Spöck, G. (1997), "Die geostatistische Berücksichtigung von a-priori Kenntnissen über die Trendfunktion und die Kovarianzfunktion aus Bayesscher, Minimax und Spektraler Sicht", Diploma Thesis, University of Klagenfurt

Kriging and Splines: Theoretical Approach to Linking Spatial Prediction Methods

Philipp Pluch

Energy and Petroleum Resources Services GmbH, Vienna, Austria
ppluch@menpet.at

1 Introduction

Spatial Statistics refers to a class of models and methods for spatial data that aim at providing quantitative descriptions of natural variables distributed in space or space and time (see Chiles and Delfiner [2], Cressie [3]). Examples for such variables are ore grades collected in a mineral field,density of trees of a certain species in a forest or CD (critical dimension) measurements in semiconductor productions. A typical problem in spatial statistics is to predict values of measurements at places where they were not observed, or if measured with error, to estimate a smooth spatial surface from the data. (Estimation of a regionalized variable.) A family of techniques, stochastic and non stochastic ones were developed in geostatistics for that interpolation problem. The general approach is to consider a class of unbiased estimators, usually linear in the observations and to find the one with minimum uncertainty, as measured by the error variance. A group of techniques, known loosely as kriging, is a popular method among different interpolation techniques developed in geostatistics by Krige [10], Matheron [11] and Journel and Huijbregts [8]. An interesting comparison of ten classes of interpolation techniques with characteristics can be found in Burrough and McDonnell [1] and in a lot of papers published recently a comparison of several interpolation techniques was made.

The goal of kriging like that of nonparametric regression is that the understanding of spatial estimation is enriched by the interpretation as smoothing estimates. On the other hand random process models are also valuable in setting uncertainty estimates for function estimates, specially in low noise situations. There are close connections between different mathematical subjects such as kriging, radial basis functions (RBF) interpolations, spline interpolations, reproducing hilbert space kernels (rhsk), PDE, Markov Random Fields (MRF) etc. A short discussion of these links is given in Horiwitz et al. [7], see Fig. 7. Splines link different fields of mathematics and statistics – and are used in statistics for spatial modelling (see more in Wahba [15]).

2 Interpolation Techniques

Interpolation techniques can be divided into techniques based on deterministic and stochastic models. The stochastic approach regards the data $\{y_i\}_{i=1}^n$ as a realisation of a random field on R^d at $t_i = \{x_{i1}, ..., x_{id}\}$, $i = 1, ..., n$ and sets $g(t)$ to be the best unbiased linear predictor of the random field at site t given the measurements.

We assume here an intrinsic random field (second-order stationary). Splines are smooth real valued functions $g(t)$. Define a roughness penalty based on the sum of integrated squared partial derivatives of a given order n. The choice of $g(t)$, which interpolates the data and minimises the roughness penalty, is known as the smoothing thin plate spline introduced by Reinsch [13].

2.1 Univariate Spline Function

A theoretical definition and historical motivation can be found in Haemmerlin and Hoffmann [5]. The natural spline $S(x) = S_n$ is a real valued function $S : [a, b] \to R$ with n knots $-\infty \leq a < x_1 < x_2 < ... < x_n \leq \infty$ with following properties

$$
\begin{aligned}
&S \in \Pi^{m-1} \text{ for } x \in [a, x_1] \text{ and } x \in [x_n, b] \\
&S \in \Pi^{2m-1} \text{ for } x \in [x_i, x_{i+1}], i = 1, ..., n-1 \\
&S \in C^{2m-2} \text{ for } x \in (-\infty, \infty) \\
&f(x_i) = f_i \text{ for } i = 1, ..., n
\end{aligned}
$$

Π^q is the class of polynomials with degree q and C^p the class of continuous functions of order p. The historical problem is to find a function f in a function space with continuos derivatives of order $(m-1)$ with

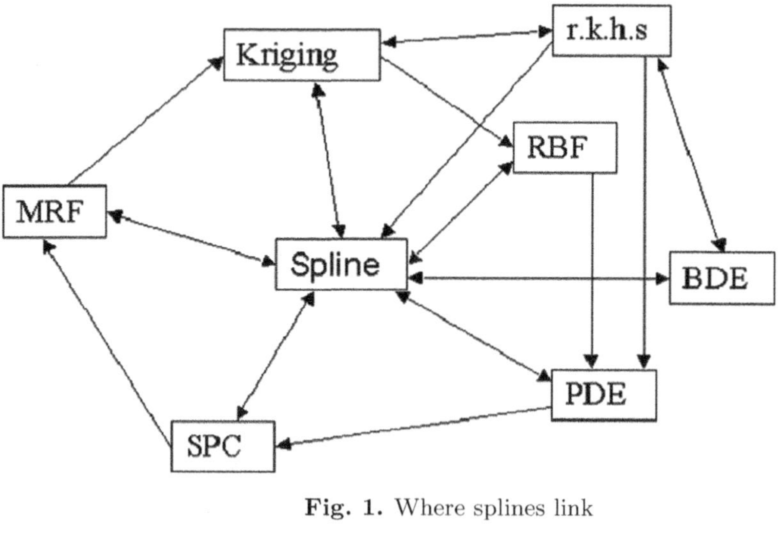

Fig. 1. Where splines link

minimum $\int_a^b (f^{(m)}(x))^2 dx$ under all functions with the property $f(x_i) = f_i$ for $i = 1, ..., n$. Schoenberg shows that the solution is the natural spline.

The statistical approach turns attention to smooth the data and not to interpolate them. The first access to this problem was done by Reinsch [13] by means of finding a function g that minimises

$$\int_{x_0}^{x_n} g''(x)^2 dx + \rho \left\{ \sum_{i=0}^{n} \left(\frac{g(x_i) - y_i}{\delta y_i}\right)^2 + z^2 - S \right\} \tag{1}$$

where z and S are auxiliary variables and ρ is a Lagrange parameter. The solution for (1) is the cubic spline:

$$g(x) = a_i + b_i(x - x_i) + c_i(x - x_i)^2 + d_i(x - x_i)^3 \text{ for } x_i \leq x \leq x_{i+1} \tag{2}$$

2.2 Multivariate Spline

In analogy with the interpolation problem in one dimension we now handle a data set recording three dimensions with a grid of points (x_i, y_i), $i = 1, ..., n$. For polynomial interpolation we can use a polynomial with low degree r of the form

$$P_r(x, y) = \sum_{p+q=0}^{r} a_{pq} x^p y^q \tag{3}$$

On an arbitrary grid it is in general not possible to construct a unique solution. Here we assume a rectangular region on which we build a rectangular $(n+1)(k+1)$ grid of the form

$$\begin{aligned} a = x_0 < x_1 < ... < x_n = b \\ c = y_0 < y_1 < ... < y_n = d, \end{aligned}$$

where in x direction and in y direction a spline will be constructed. As an analogue to the univariate B-Spline we can give the following spline base (see [5])

$$B_{1\nu\kappa} = \begin{cases} \frac{(x_{\nu+2}-x)(y_{\kappa+2}-y)}{(x_{\nu+2}-x_{\nu+1})(y_{\kappa+2}-y_{\kappa+1})} & \text{for } (x,y) \in I_1 \\ \frac{(x-x_\nu)(y_{\kappa+2}-y)}{(x_{\nu+1}-x_\nu)(y_{\kappa+2}-y_{\kappa+1})} & \text{for} (x,y) \in I_2 \\ \frac{(x-x_\nu)(y-y_\kappa)}{(x_{\nu+1}-x_\nu)(y_{\kappa+1}-y_\kappa)} & \text{for } (x,y) \in I_3 \\ \frac{(x_{\nu+2}-x)(y-y_\kappa)}{(x_{\nu+2}-x_{\nu+1})(y_{\kappa+1}-y_\kappa)} & \text{for } (x,y) \in I_4 \end{cases} \tag{4}$$

2.3 Additive Model

Now we are going to handle a $(d+1)$-dimensional data record (x_i, Y_i), with $x_i = (x_{i1}, x_{i2}, ..., x_{id})^T$; $x_1, x_2, ..., x_n$ are independent realisations of the random vector $X = (X_1, X_2, ..., X_d)$, where x_i and Y_i fulfil

$$Y_i = g(x_i) + \varepsilon_i \tag{5}$$

with $1 \leq i \leq n$, where g is an unknown smoothing function mapping from R^d to R, and $\varepsilon_1, \varepsilon_2, ..., \varepsilon_n$ are independent error terms. Our aim is to find a valid additive approximation of g by

$$g(X) \approx g_0 + \sum_{i=1}^{d} g_i(X_i) \tag{6}$$

The additive model is defined by

$$Y = \alpha + \sum_{i=1}^{d} g_i(X_i) + \varepsilon \tag{7}$$

Like in the multilinear regression case the error term is independent of X_i, $E(\varepsilon) = 0$ and $var(\varepsilon) = \sigma^2$. From (7) we can conclude that for any predictor and for all dimensions there exists at least one function g. More can be found in Hastie [6].

3 Smoothing Spline

In this paragraph we consider the following problem: We want to find a function $g(x)$ under all two times continuosly differentiable functions that minimises (1).

3.1 Univariate Approach

First we have to define:

$$h_i = x_{i+1} - x_i \text{ for } i = 1, ..., (n-1)$$

$$\Delta = \begin{cases} \Delta_{i,i} = \frac{1}{h_i} \\ \Delta_{i,i+1} = -(\frac{1}{h_i} + \frac{1}{h_{i+1}}) \text{ a}(n-2) \times n \text{ matrix} \\ \Delta_{i,i+2} = \frac{1}{h_{i+1}} \end{cases}$$

$$C = \begin{cases} C_{i-1,i} = \frac{h_i}{6} \\ C_{i,i} = \frac{1}{6}(h_i + h_{i+1}) \text{ a}(n-2) \times (n-2)\text{matrix} \\ C_{i,i-1} = \frac{h_i}{6} \end{cases}$$

With this representation the minimisation problem in (1) is equivalent to

$$\| y - g\|^2 + \rho g^T K g \tag{8}$$

where K is a quadratic penalty matrix,

$$K = \Delta^T C^{-1} \Delta.$$

Calculation of the inverse of C is possible because it is strictly diagonal dominant (see [14]). The solution to (8) is given by

$$\hat{g} = Sy \tag{9}$$

where S is the smoothing Matrix of the form

$$S = (I + \rho K)^{-1}. \tag{10}$$

Equation (10) we will later find in the solution of the kriging prediction problem and can be found also in ridge regression with natural basis and Demmler-Reinsch basis (see more in Nychka [12])

3.2 Multivariate Approach

When we use this approach with the additive model we are able to find (with help of (8)) the following result

$$g(X) \approx \sum_{i=1}^{d} \rho_i g_i^T K_i g_i + \sum_{i=1}^{d} \left(Y_i - g_0 - \sum_{j=1}^{n} g_j(x_{ij}) \right)^2 \tag{11}$$

Minimisation leads to

$$\hat{g}_l = S_l \left(y - g_0 - \sum_{j=1;\ j \neq l}^{d} \hat{g}_j \right) \tag{12}$$

$$\text{with } S_l = (I + \rho_l K_l)^{-1} \text{ for } l = 1, ..., d \tag{13}$$

S_l can be seen as smoothing matrix and K_l as penalty matrix. The following matrix equation system must be solved:

$$\begin{pmatrix} I & S_1 & S_1 & \dots & S_1 \\ S_2 & I & S_2 & \dots & S_2 \\ \vdots & \vdots & \vdots & \ddots & \vdots \\ S_d & S_d & S_d & \dots & I \end{pmatrix} \begin{pmatrix} g_1 \\ g_2 \\ \vdots \\ g_d \end{pmatrix} = \begin{pmatrix} S_1 y \\ S_2 y \\ \vdots \\ S_d y \end{pmatrix} \tag{14}$$

shortly we can write $Py = Qy$ where P is a $(nd) \times (nd)$ and Q is a $(nd) \times (nd)$ block matrix. A solution of that problem can be found by Backfitting Algorithm (see [4]). One solution to (14) is given by

$$\hat{g}(x) = \hat{Y} = g_0 + \sum_{j=1}^{d} O_j^{-1} R_j^T g_j \tag{15}$$

where R_j is a reduction matrix and O an order matrix.

3.3 Abstract Minimisation

In Wahba [15] a more general approach to the problem of spline smoothing is given. She discusses the problem of finding a function f in the Sobolev function space of the form

$$W_m : W_m[0,1] = \{f, f', f'', ..., f^{(m-1)}\text{absolutelycontinuous}, f^{(m)} \in L^2\},$$

The solution can be found by abstract analysis and abstract optimisation. Wahba gives the following equivalent smoothing problem via r.h.k.s.

$$\frac{1}{n}\sum_{i=1}^{n}(y_i - \langle \eta_i, f\rangle)^2 + \rho \parallel P_1 f \parallel_R^2 \rightarrow \min_{f \in W_m} \tag{16}$$

with the solution given by

$$\begin{aligned}
f_\rho &= \sum_{\nu=1}^{M} d_\nu \phi_\nu + \sum_{i=1}^{n} c_i \xi_i \\
\xi_i &= P_1 \eta_i, i = 1, ..., n \\
d &= (d_1, d_2, ...d_M)^T = ((T^T M^{-1} T)^{-1} T^T M^{-1}) y \\
c &= (c_1, ..., c_n)^T = M^{-1}(I - T(T^T M^{-1} T)^{-1} T^T M^{-1}) y \\
M &= \Sigma + n\rho I \\
\Sigma &= (\langle \xi_i, \xi_j \rangle), i = 1, ...n \text{ and } j = 1, ...M
\end{aligned}$$

A more general way to describe the roughness penalty function is in the form

$$J_{r+1}^d(g) = \sum_{|m|=r+1} \frac{(r+1)!}{m!(r+1-m)!} \int_{R^d} \left(\frac{\partial^{r+1} g(t)}{\partial t_1^{m_1} \cdots \partial t_d^{m_d}} \right)^2 dt \tag{17}$$

When a particular penalty is chosen then the result is invariant under rotations and translations of t.

4 Kriging

Let $\{Y(t), t \in R^d\}$ be an intrinsic or stationary random field; we additionally demand a polynomial drift of order $n > 0$. In that case the drift is a linear combination of t^m for $\mid m \mid < n$ with unknown coefficients. The subspace of the polynomial drift has an order less than n with dimension

$$\nu = \binom{d+n}{d}$$

Let

$$V_r = \begin{pmatrix} f(t_0)^T \\ F \end{pmatrix}$$
$$F = (f(t_1), ..., f(t_n))^T$$
$$EY(t) = \beta^T f(t),$$

V_r a $(n+1) \times \nu$ matrix of the drift at locations $t_0, ..., t_n$, F a $(n \times \nu)$ matrix that describes the drift at locations $t_1, ..., t_n$ and $u_{r,0}$ a vector of lentgh ν with the elements t_0^m, $\mid m \mid < n$. The covariance function is defined as $\sigma(t_i - t_j)$ for $i, j = 1, ..., n$ and $\varphi_{i,j} = \sigma(t_i - t_j)$ for $i, j = 0, ..., n$.

$$\Phi = \begin{pmatrix} \sigma^2 & \sigma_0^T \\ \sigma_0 & \Sigma \end{pmatrix}$$

where $\sigma^2 = \sigma(0)$ and σ_0 is a $(n \times 1)$ vector with elements $\sigma(t_0 - t_i), i = 1, ..., n$. Σ has the elements $\sigma_{i,j}$.

If $\{Y(t)\}$ is a stationary random field with a polynomial drift, then kriging involves predicting $Y(t_0)$ by a linear combination $\hat{Y}(t_0) = \alpha^T y$. The goal is to minimise the prediction mean squared error subject to an unbiasedness constraint.

$$var(Y(t_0) - \hat{Y}(t_0)) = var(Y(t_0) - \alpha^T y) = var(\beta^T z) = (c^T \Phi c) \quad (18)$$

under the unbiasedness constraint $Ec^T z = c^T V_r \beta = 0$. It's straightforward to minimise (18) by Lagrange multipliers to give

$$\alpha = AF_{r,0} + B\sigma_0 \quad (19)$$

where

$$A = \Sigma^{-1} F (F^T \Sigma^{-1} F)^{-1} \quad (20)$$
$$B = \Sigma^{-1} - \Sigma^{-1} F (F^T \Sigma^{-1} F)^{-1} F^T \Sigma^{-1} \quad (21)$$

for more details see Kent and Mardia [9] who also give a solution for intrinsic random fields where $\sigma(h)$ is a polynomial in h with degree $2p$. A and B can be found with the use of Moore-Penrose generalised inverse.

$$B = [(I - U_r(U_r^T U_r)^{-1} U_r^T) \Sigma (I - U_r(U_r^T U_r)^{-1} U_r^T)]^- \quad (22)$$
$$A = (I - B\Sigma) U_r (U_r^T U_R)^{-1} \quad (23)$$

5 Link Between Kriging and Thin Plate Splining

Following theorem, which identifies the kriging solution with a thin-plate spline is one of the main results in the literature (see [9]) for the discussion "kriging vs. splines".

Theorem 1. *Let $r+1 > \frac{1}{2}d$ (d is the dimension, r order of the polynomial drift) and set $\alpha = r+1-\frac{d}{2} > 0$. Then the problem of interpolating data (y_i, t_i), $i = 1, ..., n$ subject to minimising the roughness penalty (17) has a solution $g^*(t_0)$ given by*

$$g^*(t_0) = y^T A u_{r,0} + y^T B \sigma_0, \tag{24}$$

where $A, B, u_{r,0}$ and σ_0 are determined as before.

The link between kriging and thin-plate splines holds for the smoothing problem as well as for the interpolation problem shown by the last theorem. In the thin-plate spline approach a smoothing function $g(t)$ with square-integrable $(r+1)$th order derivatives has to be found such that

$$F(g, \rho) = \sum \mid y_i - g(t_i) \mid^2 + \rho J^d_{r+1}(g) \tag{25}$$

is minimised. The solution to this problem can be found in Kent and Mardia [9] or in an equivalent formulation in Nychka [12].
The optimal choice of g is given by

$$g(t_0) = y^T (I + \kappa B)^{-1} A u_0 + y^T (I + \kappa B)^{-1} B \sigma_0, \tag{26}$$

which is the same as the kriging predictor with a nugget effect.

6 Example for Kriging and Splining

Now we want to compare these two methods with the help of a modification of Wendelberger's test function (see Fig. 2), this function will be disturbed by some noise ($\sim N(0,1)$), see Fig. 3. This data set x, y and z will be smoothed

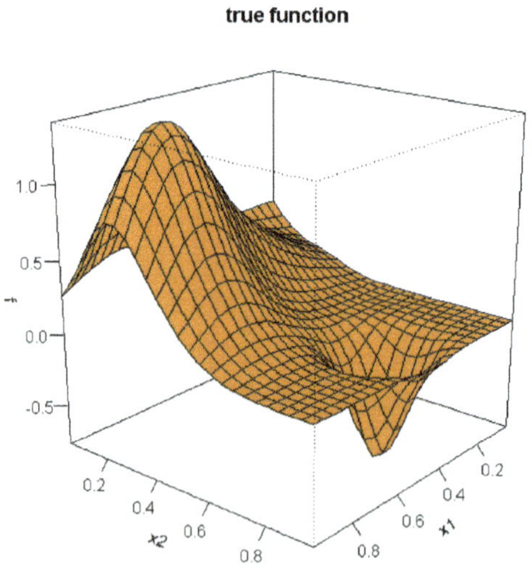

Fig. 2. Wendelberger's test function

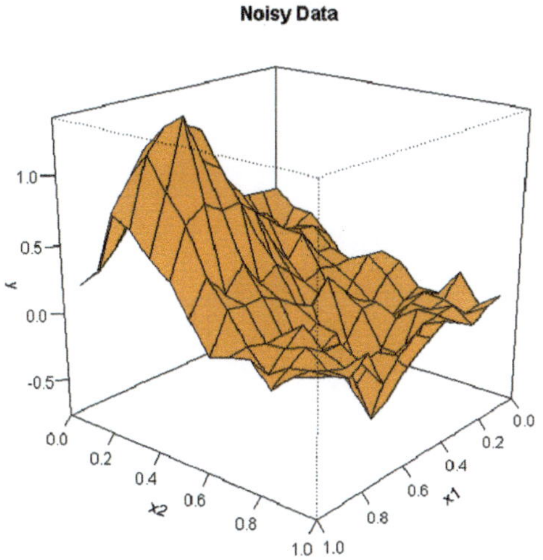

Fig. 3. Wendelberger's test function with noise

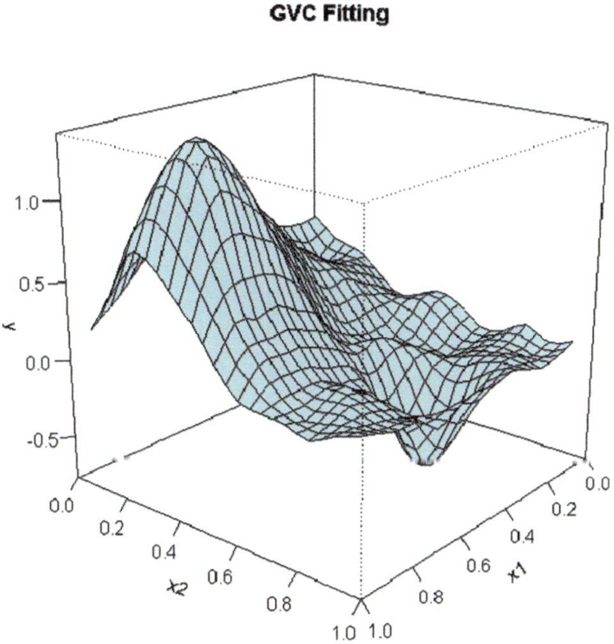

Fig. 4. GCV fitted thin plate spline

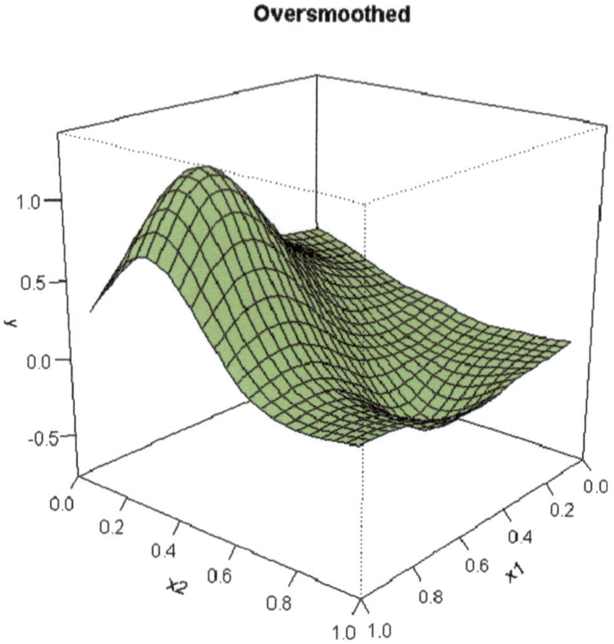

Fig. 5. Oversmoothed thin plate spline

by thin plate splines and with the help of GCV introduced in Wahba [15], see Fig. 4 and Fig. 5.

The following Fig. 6 summarises the results of GCV.

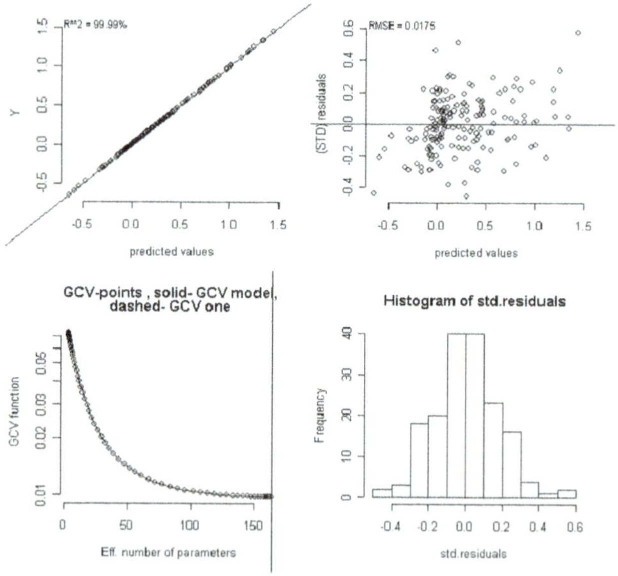

Fig. 6. Summary of kriging with matern covariance function

Finally, Fig. 7 gives a contour plot of the surface, resulting from kriging with a Matern covariance function.

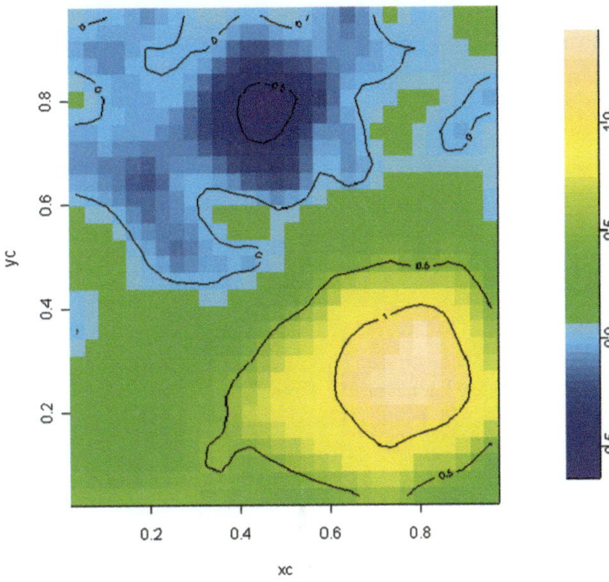

Fig. 7. Contourplot of kriging prediction

References

1. Burrough, P.A. and McDonnell, R.A. (1998) Principles of Geographical Information Systems, Oxford University Press
2. Chiles, J.P., Delfiner, P. (1999) Geostatistics, Modeling Spatial Uncertainty, Wiley, New York
3. Cressie, N.A.C. (1991) Statistics for Spatial Data, Wiley, New York
4. Green and Yandell (1985) Semi-parametric generalised linear models
5. Haemmerlin, G., Hoffmann, K.H. (1992) Numerische Mathematik, Springer Verlag, Berlin
6. Hastie, T.J., Tibshirani R.J. (1990) Generalised Additive Models, Chapman and Hall, London
7. Horiwitz et al. (1996) Fast multidimensional Interpolation. *26th Proceedings of the Applications of Computers and Operations Research in Mining Industry.*
8. Journal, A.G. and Huijbregts, Ch. J. (1978) Mining Geostatistics, Academic Press
9. Kent, J.T., Mardia K.V. (1994) The Link between Kriging and Thin-plateSpline
10. Krige, D.G. (1951) A statistical approach to some basic mine valuation problems on the Witwatersrand. *Jornal of the Chemical, Metallurgical and Mining Society of South Africa,* 52, 119–139.
11. Matheron, G. (1965) Les Variables Regionalisées et leur estimation, Mason, Paris
12. Nychka, D. (2000) Spatial Process Estimates as Smoothers. *Smoothing and Regression*, Schimeck, M. (ed.), Wiley, New York
13. Reinsch, C. (1967) and (1970) Smoothing by spline functions I, II. *Numerical Mathematics.*
14. Stoer, J. (1983) Einfuehrung in die Numerische Mathematik, Springer Verlag, Berlin
15. Wahba G. (1990) Spline Models for Observational Data, Society for Industrial and Applied Mathematics, Philadelphia, PA

ANNEX Model: Artificial Neural Networks with External Drift Environmental Data Mapping

R. Parkin[1] and M. Kanevski[2]

[1] Institute of Nuclear Safety (IBRAE), Moscow, Russia
park@ibrae.ac.ru
[2] Institute of Geomatics and Analysis of Risk, University of Lausanne, Switzerland
Mikhail.Kanevski@unil.ch

1 Introduction

The present research deals with the novel development in the field of environmental spatial data modelling with the help of Artificial Neural Networks (ANN). The following spatial prediction is considered: given the measurements of some physical quantity at finite (and relatively small) number of points, the objective is to make predictions over the considered region either on a regular dense grid (traditional mapping problem) or on irregular decision-oriented grid. In many cases in addition to the available measurements of the main variable there can be additional information: secondary variables, remote sensing images, physical model of the phenomena, soft qualitative information, etc. In the present paper a problem of spatial predictions of the primary variable using additional comprehensive information on secondary variable is considered. If there is a relationship between variables (e.g. linear correlation) the second one can be considered as an external drift. In order to solve this problem an ANNEX model (ANN + EXternal drift) is proposed. The family of ANNEX models developed for the spatial mapping problems is based on the idea of incorporation of additional spatially distributed information into the ANN as additional input(s). This secondary information is assumed to be related to the primary variable. This approach considers that additional information is available both at the training points and at all the points of the prediction grid. The similar idea traditionally is used in geostatistical "kriging with external drift" model [2]. In general, the ANNEX approach can be considered as a nonlinear modelling on a hypersurface described by input variables. In the present work the application of the ANNEX model is applied to the real case study dealing with the average long-term temperatures of air in June in Kazakh Priaralie. Additional information that will be used is the elevation at the measurement and prediction locations above the sea level. ANNEX model results are compared with the ones of both standard MLP (without extra input) and linear geostatistical estimators: kriging, co-kriging, collocated co-kriging, kriging with external drift [2].

2 ANNEX Model

The problem of spatial mapping of environmental data is rather traditional and there exist a wide variety of different prediction models to solve it. In most cases it is necessary to predict values of a spatial function (precipitation, temperatures, contamination et al.) at the unsampled points, in particular on a regular grid. Geostatistics is the well-elaborated approach to solve such problems. All geostatistical models rely on modelling of spatial correlation structures (variography) and are mainly based on a linearity hypothesis. Thus, geostatistics is a model-dependent approach: solutions highly depend on a developed model of spatial correlation. Another data driven approach is based on application of artificial neural networks [1, 3]. Neural networks are robust, nonlinear and highly flexible tools for data modelling. It was shown that ANN can be efficiently applied to spatial data modelling, especially in combination with geostatistical tools [4]. Data analysis with ANN includes several important steps: data selection and pre-processing, selection of architecture, training, testing, validation. In the present study multilayer perceptron (MLP), which is a workhorse of ANN data modelling is applied for spatial prediction of temperature. MLP being very powerful modelling tools are able to incorporate in a nonlinear manner different kinds of information and data during modelling procedure. Usually in spatial data modelling input space of ANN (independent variables) are described by geographical coordinates (e.g., x, y). Output unit (F) of ANN is a modelling function in case of univariate prediction or a vector in case of multivariate predictions. The idea of ANNEX model is as follows: if there is an additional information available at training and prediction points and related to the primary one, we can try to use it as additional inputs to the standard ANN.

Consider the examples of external information suitable for ANNEX type of modelling:

1. Availability of "cheap" information on the secondary variable(s). Consider that we are interested in a prediction of some physical quantity (primary variable) whose measurements are rather complicated and/or expensive. If there are other variables available or easily measured at all points (both measurement and prediction grids) we can try to check and to use this information in order to improve the quality of primary variable prediction.
2. Physical model of the phenomena. Consider that we are given the physical model that describes phenomena under study. To include this model into the data-driven ANNEX approach the output of the physical model at all the prediction points and at measurements locations are used as an extra input(s) for ANN. In general, secondary ANN model can be developed to model (learn) physical phenomena.

3. Physical description of the prediction region. Sometimes the additional information on the prediction region is given: altitude map, soil map, etc. Then this information can be extracted and used as an extra input to ANN. The specific case is when the remotely sensed image with useful information about the considered region is provided along with the measurements of primary variable.

The primary variable and secondary (external) information are often correlated. If the correlation is linear the methods from multivariate geostatistics can be used, e.g. cokriging. Linearity of correlation is not required by ANNEX models.

Let us remind, that the objective of advanced modelling is quite clear: by using ANNEX models we want to improve accuracy of the prediction and to reduce corresponding uncertainties, for example measured by variances.

The general question for all kind of ANNEX models is: what relationship (linear, nonlinear, stochastic or their combination) between the primary and external secondary information should be in order to make use of ANNEX efficient and how to measure these relationship? This is not an easy question in case of nonlinear and stochastic relationships between variables. The problem deals with a question of how much new information and/or new noise are introduced with ANNEX approach. An external information can dramatically change the solution in comparison with a standard model. In the present study some of these tasks are considered in an empirical way using real case study.

3 Case Study

3.1 Data Description

This case study deals with the prediction of air temperature in Kazakh Priaralie. The selected region is covering $1,400,000\,\text{km}^2$ with 400 monitoring stations. The primary variable is average long-term temperatures of air in June. Additional information that will be used as an extra ANN input is the elevation of the locations above the sea level. This information is available on a dense grid from Digital Elevation Model.

The correlation between air temperature and altitude is linear and is equal to 0.9 (Fig. 1). The linearity of correlation allowed us to use traditional geostatistical model (e.g., kriging with external drift) for modelling and comparing the results with the one obtained by ANNEX model. The similar work on modelling of air temperature applying kriging with external drift can be found in Wackernagel [7].

Following the general methodology of ANN data modelling original data were split into training and testing data sets. The spatial locations of train and test data points are presented in Fig. 2. An important and difficult problem

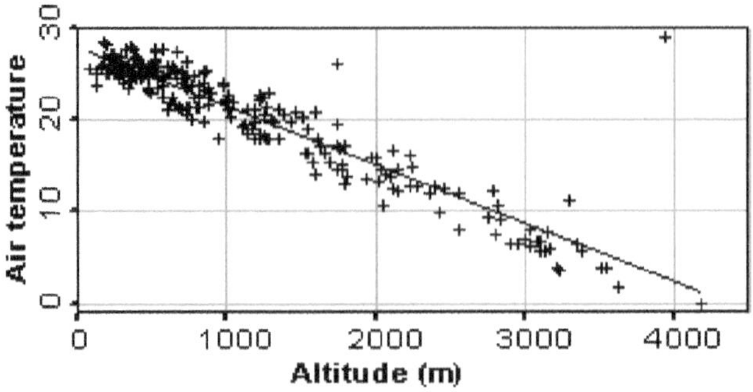

Fig. 1. The scatterplot between altitude and air temperature in June

deals with the criteria of data splitting. In most cases data are split randomly. But in case of spatial and clustered data such approach can be not adequate. In the present study the similarity of data sets was controlled by comparing summary statistics, histograms and spatial correlation structures (variograms). Since it is difficult to control both testing and training datasets, more attention was paid to the similarity of the training data set to the initial data structures of all data. Similarity of spatial structures of obtained datasets with the initial data is even more important than statistical factors. Comparison of the spatial structures was carried out with the help of variogram roses, which model anisotropic spatial correlation structures (see Fig. 3). Such comparison provided grounds that split with 168 training and 67 testing points is quite suitable for the following modelling. More advanced splitting methods can use statistical tests.

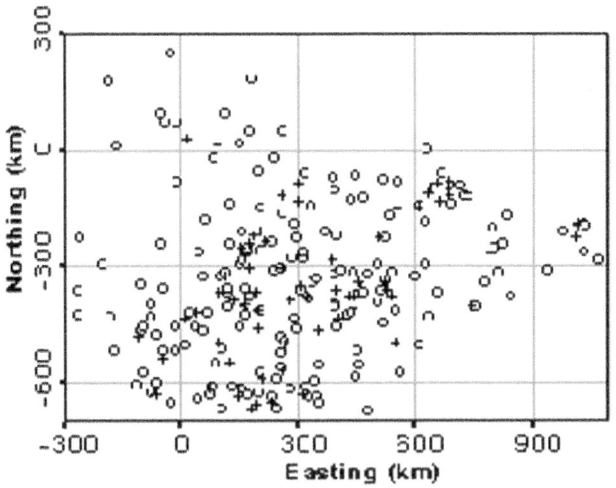

Fig. 2. The spatial location of train (*circles*) and test (*cross*) data points

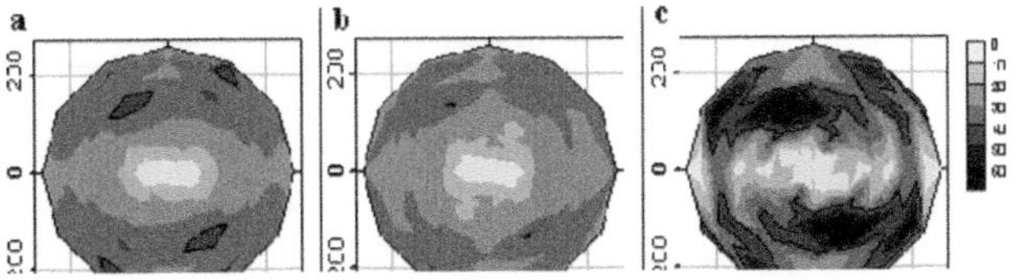

Fig. 3. Variogram roses: raw (**a**), train (**b**) and test (**c**) datasets

3.2 Geostatistical Modelling

Firstly, traditional geostatistical models were used in order to compare obtained results with ANNEX modelling results. The comparison allows understanding whether ANNEX model improves the efficiency of the prediction in this case study. The following geostatistical models were used: kriging, cokriging, collocated cokriging and kriging with external drift [2]. Results of geostatistical modelling are presented below as errors on the test data (Table 1). It can be seen that the best results among geostatistical methods on a test dataset are obtained using kriging with external drift. Cokriging results are worse than kriging ones because of screening effect [7]. Kriging and collocated kriging demonstrate analogous patterns while kriging with external drift keeps not only the large-scale structure but also small-scale variability effects ignored by kriging and cokriging models. It should be noted that results of mapping in some case should postprocessed using physical scale of the phenomena variability and in order to avoid spurious small scale structures.

Table 1. The air temperature test results for geostatistical models

model	correlation	RMSE	MAE	MRE
kriging	0.874	3.13	2.04	−0.06
cokriging	0.796	3.97	2.45	−0.11
collocated cokriging	0.881	3.05	1.95	−0.07
kriging with external drift	0.984	1.19	0.91	−0.03

3.3 ANNEX Modelling

In the present study, MLP models (as ANN) with the following parameters were used: two (traditional ANN) or three (ANNEX) input neurons, describing spatial co-ordinates (X, Y) and altitude, one hidden layer and one output neuron describing air temperature. Backpropagation training with Levenberg-Marquardt followed by conjugate gradient algorithm was used in order to avoid local minima [6]. ANN and ANNEX modelling results are presented below as errors on the test dataset (Table 2). MLP with structure 2-7-5-1 (7 and 5 neurones in the two hidden layers) showed the best result among MLPs with 2 inputs, while ANNEX model with structure 3-8-1 (8 neurones in hidden layer) gave the best result among all considered models. It is worth to mention that we used several MLP structures for the ANNEX model and found the optimum model (see Table 2). Mapping on a grid with the help of ANNEX model features similar pattern as kriging with external drift.

Table 2. The air temperature test results for ANN and ANNEX models

model	correlation	RMSE	MAE	MRE
2-7-5-1	0.917	2.57	1.96	−0.02
3-3-1	0.989	0.96	0.73	−0.01
3-5-1	0.99	0.9	0.7	−0.007
3-7-1	0.991	0.85	0.66	−0.004
3-8-1	0.991	0.84	0.68	−0.001
3-9-1	0.991	0.88	0.69	−0.01
3-10-1	0.99	0.92	0.74	−0.01

3.4 Noise Effect

As it was mentioned above, an important problem concerns the question of the quality of additional data: there is a dilemma between introducing new information and/or new noise. In order to test the influence of noise on models (1) we have generated noise which was used as an additional input; (2) external information (elevation) was contaminated by noise. In fact, the objective was to check the robustness and the stability of solution obtained with different models. Firstly, we have considered 100% noise injection. This model is similar to the models of noise injection into hidden layer. The noise was modelled in the following way: the altitude values are remained the same (distribution is the same), while the spatial coordinates interchanged randomly (spatial structure is reduced to a nugget model). Such procedure is well known in time series modelling: distribution of data is preserved while correlation in time is destroyed. In the same manner way we changed only 10% of points and examined the obtained results (see Table 3). It can be seen that presented ANNEX model is unsusceptible to noise in external information.

Table 3. The air temperature noise test results

model	correlation	RMSE	MAE	MRE
kriging	0.874	3.13	2.04	−0.06
kriging external drift	0.984	1.19	0.91	−0.03
3-8-1	0.991	0.84	0.68	−0.001
3-8-1 (1000.839	3.54	2.37	−0.13	
3-8-1 (100.939	2.32	-1.49	−0.003	
kriging – ext. drift (100.941	2.23	1.54	−0.06	
3-8-1 (100.899	2.81	1.52	−0.08	
kriging – ext. drift (100.903	2.81	1.59	−0.103	

4 Conclusions

An Artificial Neural Networks with External drift (ANNEX) model for the analysis and mapping of spatially distributed data was applied to the real data. It was shown that additional spatially distributed information can be efficiently used by ANNEX and gives rise to better analysis and modelling of environmental data. Promising results presented are based on the real case study of air temperature mapping. Other kinds of Machine Leaning models (besides ANN) can be used with possible modifications in the proposed framework. The advantage of the ANNEX model is its ability to model any nonlinear relationships between variables. An interesting feature found in the study is robustness and stability of the ANNEX solution versus noise. This problem should be studied in more detail. ANNEX model performed better even in the case of linear correlation between primary and secondary information that is favourable to kriging with external drift. An even more interesting study should consider nonlinear relationships between data and external information.

Acknowledgements

The analysis and presentation of the results as well as MLP and geostatistical modelling were performed with the help of GEOSTAT OFFICE software ([5], http://www.ibrae.ac.ru/ mkanev). The work was supported by INTAS Aral Sea project #72. Authors thank S. Chernov and V. Timonin for programming GEOSTAT OFFICE software, which was extensively used in the research.

References

1. Bishop CM (1995) Neural Networks for Pattern Recognition, Clarendon Press, Oxford
2. Deutsch CV, Journel AG (1998) GSLIB Geostatistical Software Library and User's Guide, Oxford University Press, New York, Oxford
3. Haykin S (1999) Neural Networks, A Comprehensive Foundation, Second Edition, Prentice Hall International, Inc., Prentice Hall
4. Kanevski M, Arutyunyan R, Bolshov L, Demyanov V, Maignan M (1996) Artificial neural networks and spatial estimations of Chernobyl fallout. Geoinformatics, vol. 7, pp. 5–11
5. Kanevski M, Demyanov V, Chernov S, Savelieva E, Serov A, Timonin V (1999) Geostat Office for Environmental and Pollution Spatial Data Analysis. Mathematische Geologie band 3 April 73–83, CPress Publishing House, Dresden
6. Masters T (1995) Advanced Algorithms for Neural Networks. A C++ Sourcebook, John Wiley & Sons, Inc., New York
7. Wackernagel H (1995) Multivariate Geostatistics, Springler-Verlag, Berlin

Regional Classification of Indoor Radon Data with Support Vector Machines and Geostatistical Tools

A. Chaouch[1], M. Kanevski[1], M. Maignan[1], A. Pozdnoukhov[2], J. Rodriguez[3], and G. Piller[3]

[1] Institute of Mineralogy and Geochemistry, University of Lausanne, Lausanne, Switzerland aziz.chaouch@etu.unil.ch
[2] Institute of Geomatics and Analysis of Risk, University of Lausanne, Switzerland Alexei.Pozdnoukhov@unil.ch
[3] Swiss Federal Office of Public Health (OFSP), Bern, Switzerland

1 Introduction

For the last 20 years, the Swiss Federal Office of Public Health (OFSP) has performed more than 65,000 indoor radon measurements throughout whole Switzerland. Swiss Indoor radon data are noisy and poorly spatially correlated. They feature large low-scale variability and a strong spatial clustering. Univariate distribution is positively skewed and heavy-tailed. Thus one possible way to deal with these data is to transform them into indicators relative to a decision threshold and apply spatial statistics to these indicators. Indeed when considering decision making, the task is often to classify indoor radon data into low or high concentration level. This kind of two-class classification task is commonly solved by geostatistical interpolations of indicators using kriging and/or conditional simulations. However, geostatistical approaches depend on several assumptions about the data (i.e. stationarity) and require modelling of the variogram (see Chiles and Delfiner [1]), a task that is, while sometimes possible, often very difficult and time consuming with indoor radon data. In consequence, data-driven approaches such as support vector machines (SVM) are considered as an alternative to geostatistical approaches. In this paper, their performance in application to indoor radon data classification is assessed in comparison the one of indicator kriging (IK) and sequential indicator simulations (SIS).

2 Data Pre-processing

This study will focus on a small square of 25 × 25 km located at the north-eastern end of Switzerland, near the "Bodensee". Over this region, stationarity might be assumed and the reasonable amount of data allows us to perform a comprehensive spatial analysis with both geostatistical methods and SVM. Selection of continuous radon levels recorded in ground floor of inhabited houses was performed.

2.1 Indicator Transform

Continuous indoor radon levels are transformed into discrete binary indicators [0;1] for geostatistical methods or $[-1;+1]$ for SVM relative to a user-defined threshold. In the present study, the threshold is set at $45\,\mathrm{Bq/m^3}$, close to the median of the regional distribution of indoor radon levels. That level was chosen for the present methodological study and does not reflect decision level defined by Swiss federal law [6]. Indicator I at location u is built by comparing the local indoor radon level $Z(u)$ to the decision threshold Z as follow:

$$I(u;Z) = 1\ \text{(geostat) or}\ -1(SVM)\ \text{if}\ Z(u) < Z; Z = 45\,\mathrm{Bq/m^3}$$
$$I(u;Z) = 0\ \text{(geostat) or}\ +1(SVM)\ \text{otherwise}$$

After binary coding of the data, the dataset contains 658 indicators. To assess classification abilities of methods, a subset of 158 data is kept for validation purposes only, leaving 500 source data available for the spatial analysis, see Fig. 1.

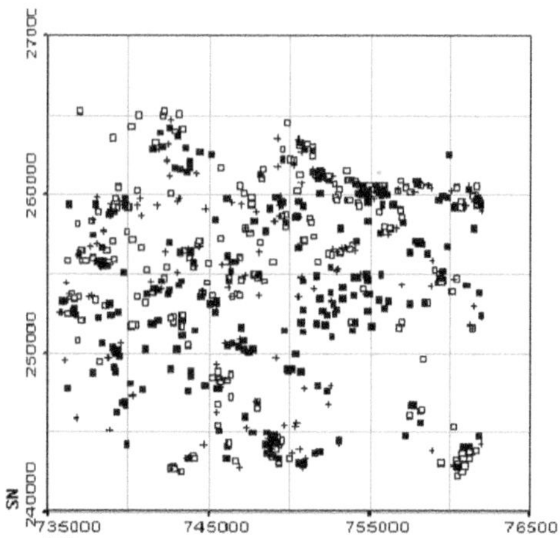

Fig. 1. Plot of indicator 1 (*white box*), indicator 0 (*black box*) and validation data (*marks*)

2.2 Indicator Declustering

Obvious spatial clustering of source indicators induces a bias on univariate statistics that are no more representative for the entire studied area. Morishita diagram [4], measuring degree of spatial clustering at different scales, shows that at any scale indicators 1 are slightly more clustered than indicators 0. This preferential clustering tends to overestimate Indicator mean value. In order to retrieve representative univariate statistics from the entire area, Cell declustering and Voronoi polygon declustering [3] were used to compute the weighted statistics of data, see Table 1.

Table 1. Univariate statistics of raw and declustered indicators

	raw indicators	cell-declustering	polygon declustering
mean	0.508	0.462	0.477
stDev	0.500	0.499	0.499
nb of data	500	500	500

Average value of declustered mean (0.470) obtained with both declustering techniques is assumed to be the representative indicator mean value over the entire area of investigation.

3 Geostatistical Tools

When dealing with indicator transformed data [0;1], geostatistical tools such as kriging and simulations provide an estimate of the probability that indicator 1 prevails at any unknown location u. This estimation requires definition and modelling of the variogram of indicators.

3.1 Exploratory Variography of Indicator Data

The interpolation is based on the spatial structure of indicator data I which is described and modelled by the semi-variogram (1) or variogram function γ where h is the lag of the variogram (i.e. the Euclidean distance between pairs of N points considered for the calculation).

$$\gamma_1(h) = \frac{1}{2N}\sum_{i=1}^{N}(I(u_i) - I(u_i + h))^2 \tag{1}$$

Spatial structure of indicator data for threshold $45\,\mathrm{Bq/m^3}$ is weak but existent. For improved visibility and easier modelling, common lag tolerance of half the lag (traditional variogram) is here increased to three times the lag. Resulting variogram shape is smooth and easier to model (regularised variogram).

Correlation distance of indicator data is close to 2 km with a weak anisotropy being present along NNW-SSE direction, see Fig. 3. High variability of indicator is obvious with nugget accounting up to 70% for the total variability that is a reason of low classification efficiency.

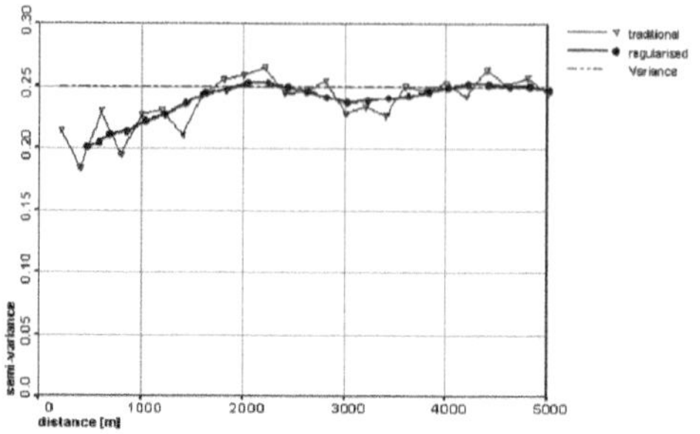

Fig. 2. Traditional and regularized omnidirectional variogram

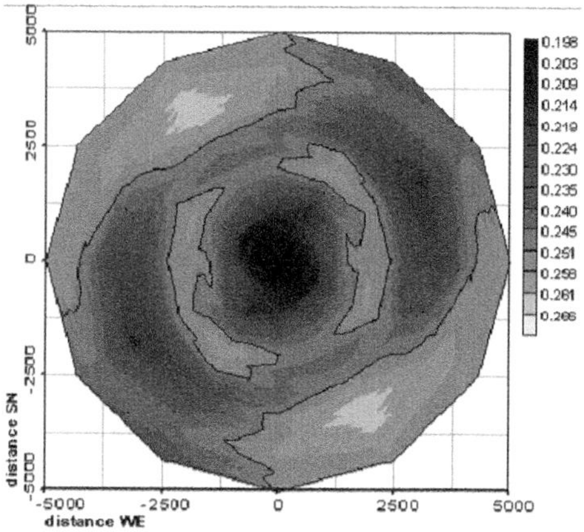

Fig. 3. Variogram rose

3.2 Indicator (Simple) Kriging

Consider the problem of estimating the indicator value $I(u;Z)$ with known constant mean m at any location u using N available hard indicators I_k defined at the threshold Z. The indicator kriging estimate is a linear combination of available indicators.

$$I(u;Z) = \sum_{k=1}^{N} \lambda_k \cdot I_k(x_k;Z) + \left[1 - \sum_{k=1}^{N} \lambda_k\right] \cdot m \tag{2}$$

Weights λ_k assigned to available indicators I_k aimed at minimizing the error variance and thus are provided by the simple kriging system:

$$\sum_{k=1}^{N} \lambda_k \cdot C_1(x_j, c_k) = C_1(x_j, u); j = 1, \ldots, N \tag{3}$$

where $C_i(a,b)$ is the covariance function of indicators between locations a and b. In case of second order stationarity the covariance function is linked to the variogram function (1) of indicators $\gamma_I(h)$ where h is the distance between locations a and b and σ_I^2 is the expected variance of indicators.

$$\gamma_I(h) = \sigma_I^2 - C_I(a, b) \tag{4}$$

When available indicators are coded either as 0 or 1 relative to threshold Z, the estimated indicator value can be seen as the probability that indicator 1 prevails at location u. It is then possible to decide on which probability level the distinction between class 0 and class 1 is achieved. In this study, declustered mean value was used as simple kriging mean m and the distinction between classes is performed as follow:

$$I(u; Z) = 1 \text{ if } I\ (u; Z) > 0.5$$
$$I(u; Z) = 0 \text{ otherwise}$$

3.3 Sequential Indicator Simulations

Smoothing effect of kriging induced by the error minimization is well documented [4]. Comprehensive analysis of indicator response requires full reproduction of variability and thus sequential simulations are to be considered.

The key problem is here to rebuild both the distribution (i.e. proportions) and the spatial structure (i.e. variogram) of available indicators still by conditioning them using kriging. The task is performed as follow:

- choose a random path visiting each node n_i of the interpolation grid
- perform simple kriging interpolation of available indicators at first node n_1 of the path. As defined in indicator simple kriging, this gives the probability that $I(n_1; Z) = 1$.
- draw a random indicator [0;1] from probability function $I(n_1; Z)$
- add the simulated indicator to the dataset and consider it as a new data
- proceed to next node until the interpolation grid is filled.
- restart the whole process many times using different random paths to build many simulated images

Simulated images are all equally-probable and each one provide a different possible reality of indicator configuration that honours both raw data and their spatial structure, see Fig. 4.

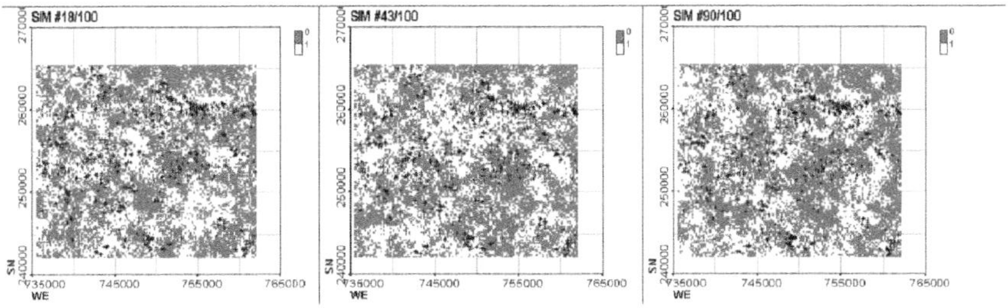

Fig. 4. Some of equally-probable single realizations of the simulation process

From the number of different simulated images, it is possible to infer the probability to have indicator 1 at any location. As for indicator kriging, final classification map is produced by choosing probability 0.5 as the decision level:

$$I(u; Z) = 1 \text{ if } Z(u) < Z$$
$$I(u; Z) = 0 \text{ otherwise}$$

4 Support Vector Machines (SVM) for Classification

SVM is a machine learning classification algorithm able to classify data into a binary format $[-1; 1]$ relative to a user-defined threshold by building a maximum margin hyperplane separating the two classes, see Kanevski et al. [5].

4.1 Statistical Learning Theory

Support Vector Machines is based on statistical learning theory defined by Vapnik-Chervonenkis [7]. Statistical learning theory studies the process of inferring based only on a finite amount of data. The fundamental problem is thus called generalization. Best generalization abilities are retrieved with a model of limited complexity. Indeed, simplistic models are not able to extract enough information out of the data while very complex models tend to overfit data and thus provide poor generalization performances. Therefore, SVM is based on the structural risk minimisation principle, aiming to minimize both the empirical risk (testing error) and the complexity of the model.

4.2 Linear Classification

Consider the classification problem of linearly separable training data labelled $-1; 1$ relative to the decision threshold by a hyperplane $H : w \cdot z + b = 0$ where z is the input of training data (namely coordinates). Label y_i of any point i is defined by looking on which side of H point i is lying. Formally, this yields to evaluation of (5).

$$y_i = sign(w \cdot z_i + b) \tag{5}$$

Several hyperplanes are able to separate these data but the unique optimal solution considers a maximum margin hyperplane H where the distance δ between all points z and H is maximum. It can be shown that maximizing this distance yields to minimizing the norm of the weight vector w. The maximum margin hyperplane should satisfy the condition that training data are correctly classified:

$$y_i \cdot (w \cdot z_i + b) \geq 1 \forall i \tag{6}$$

This leads to an optimisation problem under constraints that can be solved by introducing positive Lagrange multipliers α.

$$L(w, z, \alpha) = \frac{1}{2} \parallel w \parallel^2 - \sum_{i=1}^{N} \alpha_i (y_i \cdot (w \cdot z_i + b) - 1) \tag{7}$$

$L(w, z, \alpha)$ must be minimized with respect to w and b. Applying the optimality conditions yields to 2 solutions:

$$\frac{\partial L}{\partial w} \Longrightarrow w = \sum_{i=1}^{N} \alpha_i \cdot y_i \cdot z_i \tag{8}$$

$$\frac{\partial L}{\partial b} = 0 \Rightarrow \sum_{i=1}^{N} \alpha_i \cdot y_i = 0 \tag{9}$$

Equation (6) shows that the weight vector w and thus the hyperplane H is defined only in terms of data with associated $\alpha > 0$. These data are called support vectors. Attribute b of the hyperplane is chosen so that it maximizes the margin.

Classification of any new point is then performed by substituting (8) in (5), what gives the decision function:

$$y_{new} = sign \left(\sum_{i=1}^{\#SV} \alpha_i y_i z_i^{SV} z_{new} + b \right) \tag{10}$$

In the linear case, the classification is thus achieved by solving the dot product between support vector inputs and inputs of the new point to be classified. Inputs are commonly spatial coordinates of points but they may also contain additional information.

4.3 Non Linear Classification

Following Cover's theorem, non linear classification of data can be achieved by mapping the data into a higher dimensional feature space where data becomes linearly separable [7]. As the dimension of the feature space where linear separation is possible is unknown, kernel functions Script K are introduced. They are able to solve the non linear classification problem in the input space without explicitly going in the feature space where data would become linearly separable. This is the so-called "Kernel trick".

Eligible kernels should satisfy Mercer theorem [7, p. 423] and can take various forms. In this study, the Radial Basis Function Kernel (11) with standard deviation σ (kernel width) has been used for SVM classification.

$$\mathcal{K}(z_i, z_j) = \exp\left(- \frac{\parallel z_i - z_j \parallel}{2\sigma^2} \right) \tag{11}$$

4.4 Soft Margin Classifier

When dealing with noisy data such as indoor radon levels, it is not always advisable to correctly classify all training data. Indeed, data are likely to contain errors or incoherent values. The classifier is then built to allow misclassification of points that have a too strong impact on the boundary definition in order to improve generalisation abilities. The resulting boundary has a so-called "soft margin". Implementation of this additional constraint on the optimisation problem is done by introducing a new parameter called C-value in addition to slack variables ξ_i. Basically misclassified points i are on the wrong side of their margin by an amount $C \cdot \xi_i$ [2]. The Lagrangian can then be rewritten as follow:

$$
\begin{aligned}
&L = \tfrac{1}{2} \parallel w \parallel^2 + C \cdot \sum_{i=1}^{N} \xi_i - \sum_{i=1}^{N} \alpha[y_i \cdot (w \cdot z_i + b) - (1 - \xi_i)] - \sum_{i=1}^{N} \mu_i \xi_i \\
&\xi \geq 0, i = 1, \ldots, N \\
&0 \leq \alpha_i \leq C
\end{aligned} \tag{12}
$$

C value is hence a trade-off between margin maximisation and classification error.

4.5 SVM Parameters Tuning

SVM parameters such as kernel width and C value are unknown and must be tuned. Therefore, source dataset is split into training and testing data. Model is built with training data and parameters are tuned according to testing data using different combinations of kernel widths and C values. Optimality of SVM parameters is reached following the structural risk minimization principle [7]. Testing error and complexity of the model that is defined in term of number of support vectors are both considered. The more support vectors, the more complex the model.

Ten different random splits of 400 training and 100 testing data were performed to tune SVM parameters. Indeed optimal parameters may vary for different random splits due to noisy data and/or possible algorithmic instabilities. Average values of both testing error and normalized number of support vectors for the ten random splits are presented on a map.

As suggested on Fig. 5, there is no clear unique solution. Minimal test error lies on a straight line pointing out a fairly linear dependence between optimal kernel width and logarithm of C value. However, two white patches of low testing error can be seen on this line, one for a kernel width of 1000 m and the other at approximately 3 km. The solution with kernel width of 3 km and log(C) of 3 can be built by less support vectors (Fig. 6) and is therefore preferred over the other. Final classification with optimal parameters is applied to training and testing data altogether. Under such configuration, the final classification is performed using only 341 support vectors out of the 500 original data.

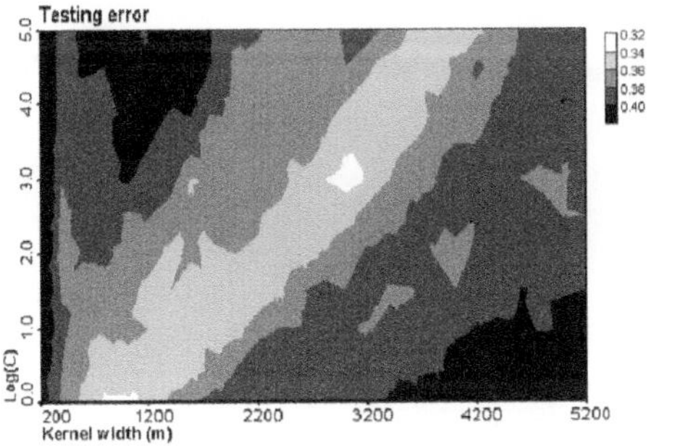

Fig. 5. Average classification error on testing data (map of empirical risk)

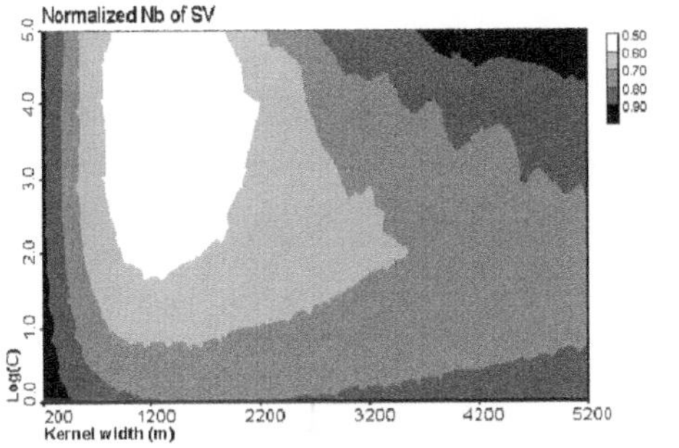

Fig. 6. Average normalized number of support vectors (map of model complexity)

5 Classification Results

Classification maps produced using both geostatistical methods and support vector machines are presented in this section. Interpolation grid has square cell-size of 200 m and present postplot of source (marks) and validation data (circles).

Classifications obtained with geostatistical methods are very similar as they use the same model of spatial structure: the variogram. However, SIS

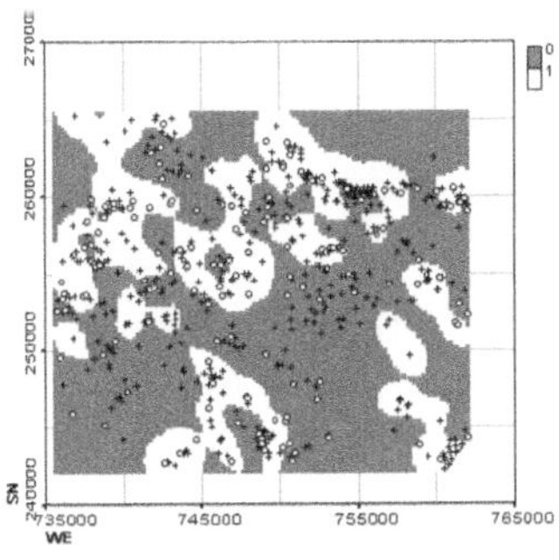

Fig. 7. IK classification

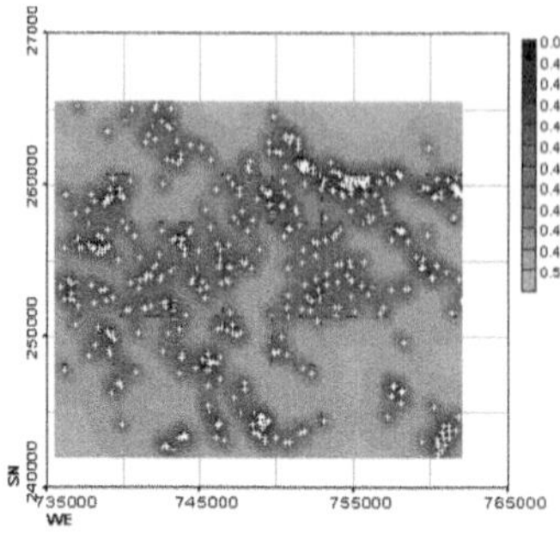

Fig. 8. IK standard deviation

classification is noisier as its purpose is to rebuild variability of data while IK aim at minimizing error variance thus smoothing out resulting figures. SVM classification produced with optimal parameters show large scale smooth delimiting contours that feature similar patterns to the one obtained with SIS and IK, see Figs. 7–9. However, it is obvious that SVM classification provides less 0 indicators than other approaches. Unmatched classifications are mainly located in poorly sampled regions where geostatistical estimates are less accurate like in the limits of the area (Fig. 10).

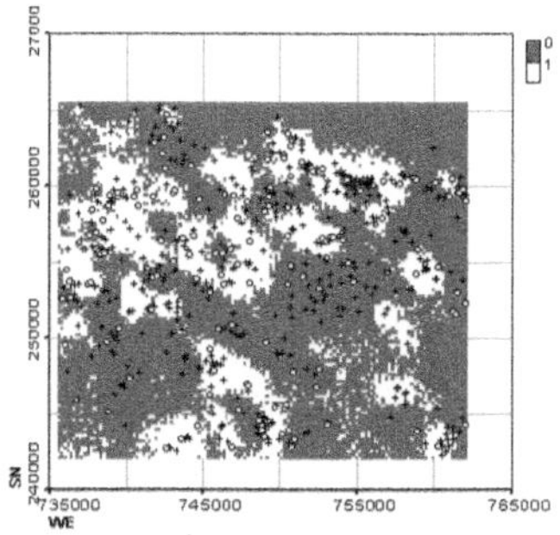

Fig. 9. SIS classification

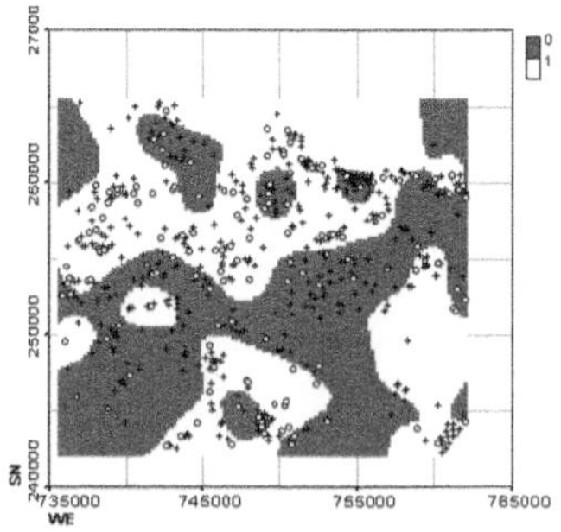

Fig. 10. SVM classification

Table 2. Validation of classification

classification method	% of misclassified validation data
IK	30.38
SIS	29.75
SVM	32.91

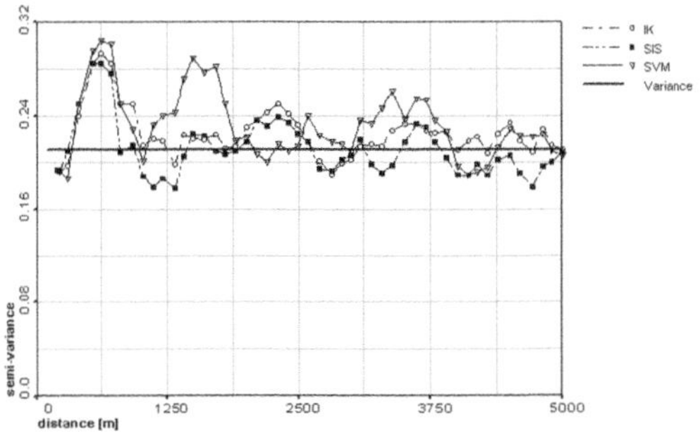

Fig. 11. Omnidirectional regularized variograms of validation residuals

Accuracy of classification results is roughly assessed using validation data (Table 2). Misclassification of validation data is rather similar for all reviewed methods and stands around 30%. This poor predictability is explained by the high variability of indicators as suggested by the elevated nugget seen on the indicator variogram (Fig. 2). Apparently SVM cannot perform better than geostatistical methods under such configuration. An explanation for this might be found in the design of the validation experiment itself. As SVM are able to misclassify some data according to their direct influence on the boundary, validation data that contain as much noise and/or errors than source data shouldn't be necessarily correctly classified. Another validation experiment consists then to apply exploratory variography on validation residuals, see Fig. 11. Resulting variograms are fluctuating around sill and suggests that all reviewed methods despite their different nature are able to extract the weak spatial information out of the data.

6 Conclusions

Classification abilities of geostatistical approaches and SVM for a median value test decision threshold ($45\,\mathrm{Bq/m^3}$) are presented in this study. Despite their poor classification abilities that are easily explained by the high variability of indoor radon indicator data, all reviewed methods are able to efficiently extract spatial information out of the data. In particular SVM are promising for indoor radon binary classification as they don't require any prior assumption on data such as stationarity. They may therefore be applied to classification problem over large regions if not over the whole country. A specific feature of SVM is its ability to consider data as noise while geostatistical approaches force conditioning of data by kriging. SVM were then able to achieve same classification abilities as geostatistical approaches with potentially less data (support vectors). Despite the requirement of many different random splits of training and testing data to assess convergence of optimal solution, the main advantage of SVM is to avoid the difficult and time-consuming task of variogram definition and modelling which is known to be very tricky in case of indoor radon data. Moreover SVM may incorporate additional data/knowledge relevant to radon data classification (geology, soil permeability, etc.) as added inputs to spatial coordinates.

Acknowledgments

The work was supported in part by the INTAS Aral sea grant #72.

References

1. Chilés J.P and Delfines P (1999) Geostatistics, Modelling Spatial Uncertainty. Wiley-Interscience Publication, New York
2. Hastie T, Tibshirani R and Friedman J (2001) The elements of Statistical Learning. Data mining, inference and prediction. Springer-Verlag, Berlin
3. Isaaks E. H and Srivastava R. M (1989) An Introduction to Applied Geostatistics. Oxford University Press, New York
4. Kanevski M and Maignan M (2004) Analysis and modelling of spatial environmental data. Presses polytechniques et universitaires romandes (ISBN 2-88074-507-1)
5. Kanevski M, Pozdnukhov A, Canu S, Maignan M, Wong P and Shibli S (2001) Support vector machines for classification and mapping of reservoir data. Soft Computing for Reservoir Characterization and Modeling. Springer-Verlag, Berlin, pp. 531–558
6. OFSP (2002) Information on radon http://www.bag.admin.ch/strahlen/ionisant/radon/f/
7. Vapnik V.N (1998) Statistical Learning Theory. Wiley-Interscience Publication, New York

Daubechies Wavelets for Identification of Rock Variants from IR Spectra

Vera Hofer[1], Jürgen Pilz[2], and Thorgeir S. Helgason[3]

[1] Department of Statistics and Operations Research, Karl-Franzens University Graz, Graz, Austria
vera.hofer@uni-graz.at
[2] Department of Statistics, University of Klagenfurt, Klagenfurt, Austria
juergen.pilz@uni-klu.ac.at
[3] Petromodel Ltd, Reykjavik, Iceland
thorgeir.helgason@petromodel.is

1 Introduction to the Data and the Problem

Aggregates, i.e. sand, gravel and crushed rock, are the most frequently used construction materials worldwide, i.g. in concrete, cement, asphalt etc. Among igneous rocks, granite and basalt are the most important. The properties of all construction materials need to be appropriate for their intended purpose. Some applications leave room for choice among many different rock variants, others require a thorough inspection of the particular features of the rocks. The question of whether the rock will resist physical and chemical loads, is of particular importance. Specific imperfections in granite result from the transformation of feldspar to kaolinite, or the decay of biotite and may lead to reduced strength [6]. As rocks are increasingly being used up to the limits of their mechanical strength, material tolerances are decreasing. This leads to the demand for ever more careful assessment of rocks. In order to decrease the costs of damages arising from improper use of aggregates, and to substantially reduce production costs, the aggregates industry is interested in effective quality control.

This calls for an efficient and fast method for classification of aggregates. Automatic means for identifying suitable rock characteristics and their variation so have to be devised. It is well known that matter treated with light of different wavelengths shows characteristic features that are suitable for qualitative and quantitative analysis and therefore for identification of substances (e.g. [9]). The optical characteristics of the material investigated is expressed in a spectrum, i.e. a plot of the absorption, transmission, reflection, or emission intensity as a function of wavelength, frequency, or energy [2].

Theoretical studies show that different substances have characteristic spectra in certain wavelength regions, even though the appearance of these spectra may vary considerably, depending on the parameters used. This raises the question, whether a statistical method for classification of aggregates is possible.

The aim of the EUREKA project, PETROSCOPE, is to develop an automatic testing instrument for process and quality control in the construction aggregates industry [1]. In this paper, stemming from the PETROSCOPE project, the identification of two types or variants of granite, called Granite 1 and Granite 2, by means of their reflectivity in mid-infrared light is investigated. Ten samples, i.e. particles of size 16–32 mm for each of the two types of Finnish granite, supplied by Lohja Rudus Oy from Hiiskula gravel pit, were collected by the Geological Survey of Finland. Then the samples were irradiated with infrared light at equidistant wavenumbers from 560 to 4000 cm^{-1} and from three positions for each particle. These measurements, performed at VTT Electronics in Finland, resulted in three curves per particle or sample and therefore 60 curves all together [15]. Figure 1 shows the spectral lines of the samples for each of the two granite classes or variants.

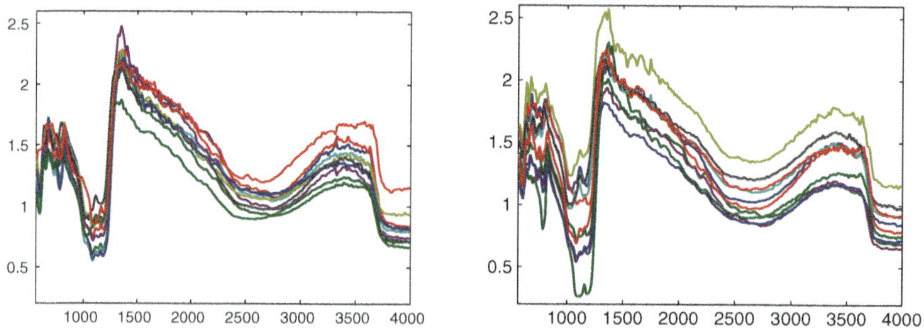

Fig. 1. Reflectivity of two types of granite aggregates in infrared between wavenumber 560 nm bis 4000 nm, first measurement or position for each sample

In general, the spectral lines of the curves seem to be very similar, although there are also some specific characteristics. This impression is mirrored by the mean curves of the classes in Fig. 2, which raises the statistical question, whether the differences in the shape of the curves are systematic or just random.

The data observed are continous curves not single observations of scalars [7, 8], even though the curves were measured at discrete knots and therefore represented by data vectors $\mathbf{x}_i = (x_{i1}, \ldots, x_{in})$, where n indicates the dimension of the vectors, i.e. the number of observation knots. In statistical problems dealing with spectra, the high dimensionality of the data causes problems in applying the common techniques of multivariate statistics because the number of samples compared to the number of observation knots is very small. In this study the proportion of samples to predictors is about 1:8. Even partial least

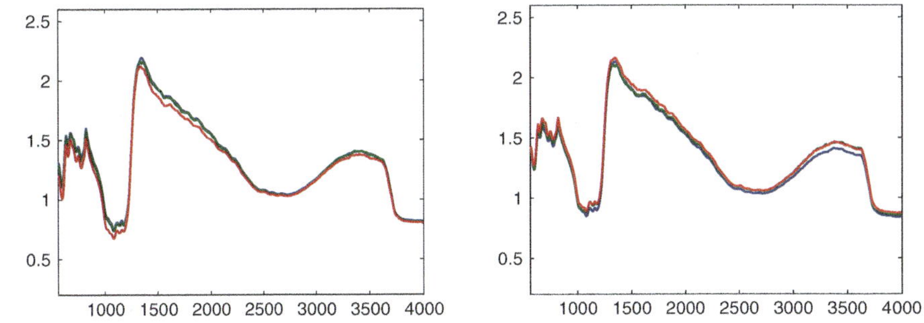

Fig. 2. Mean curves from each of three measurements of reflectivity for ten particles or samples in both classes of granite

squares (PLS) does not lead to reasonable classification error rates because of the similarity of the curves, as discussed in Sect. 2.

Several techniques exist in order to overcome the problem of high dimensionality (multicollinearity). In the following section feature reduction by a basis expansion is described. We use the fact that the curves observed are in $L^2(\mathbb{R})$, which is a Hilbert space. The idea is to choose a proper basis and then consider a subspace that contains the essential information of the signal. This information is received by an orthogonal projection of the signal onto this subspace.

In the current research we use a wavelet basis. Wavelets have adequate local properties and have turned out to be appropriate for statistical modelling of high dimensional data, as the characteristic features are summarized by a few basis coefficients.

1.1 Wavelets

Wavelets are functions that are received from a mother wavelet $\psi \in L^1(\mathbb{R}) \cap L^2(\mathbb{R})$, which is ([13]):

(i) "waving" above and below the abscissa, i.e. $\int_{-\infty}^{\infty} \psi(t)\, dt = 0$. This means that ψ has mean zero,
(ii) well localized.

By translation and dilation of the mother wavelet ψ we obtain the family

$$\psi_{\mathrm{j\,k}}(x) = 2^{\frac{\mathrm{j}}{2}}\, \psi(2^{\mathrm{j}}\, x - k) \,, \tag{1}$$

where the integers j and k are the spatial parameters of the discret wavelet transform. Two technical terms are usual in signal processing: The factor of stretch or compression is called scale. The inverse of scale is the resolution. The higher the resolution the better the approximation and the lower the scale. The relation among level j, resolution and scale is shown in Table 1. (In this paper Mallat's indexing is used [13].)

Table 1. Relation between level, scale and resolution

level	-2	-1	0	1	2
scale	4	2	1	$\frac{1}{2}$	$\frac{1}{4}$
resolution	$\frac{1}{4}$	$\frac{1}{2}$	1	2	4

Under mild conditions on ψ the family (1) constitutes an orthonormal basis of the $L^2(\mathbb{R})$ so that each function $f \in L^2(\mathbb{R})$ can be expressed as

$$f = \sum_{\mathrm{j,\,k} \in \mathbf{Z}} d_{\mathrm{j\,k}}\, \psi_{\mathrm{j\,k}} \,.$$

To get a decomposition of a signal from low to high resolution, we define a scaling function for a fixed level $\mathrm{J_0}$, the father wavelet ϕ, so that the functions

$$\phi_{\mathrm{J_0\,k}} = 2^{\frac{\mathrm{J_0}}{2}}\, \phi(2^{\mathrm{J_0}} x - k) \qquad k \in \mathbf{Z}$$

constitute an orthonormal basis of

$$V_{\mathrm{J_0}} = span\,(\{\phi_{\mathrm{J_0\,k}} \,|\, k \in \mathbf{Z}\}) = span\,(\{\psi_{\mathrm{j\,k}} \,|\, j < J_0 \wedge k \in \mathbf{Z}\}) \,.$$

So we get a new orthonormal basis of the $L^2(\mathbb{R})$

$$\{\phi_{\mathrm{J_0\,k}},\, \psi_{\mathrm{j\,k}} \,|\, j \geq J_0 \wedge k \in \mathbf{Z}\} \,.$$

For a fixed level j define

$$W_{\mathrm{j}} = span\,(\{\psi_{\mathrm{j\,k}} \mid k \in \mathbf{Z}\}) \,.$$

Then it can be seen easily that the space V_{j+1} is decomposed into two orthonormal subspaces

$$V_{\mathrm{j}+1} = V_{\mathrm{j}} \oplus W_{\mathrm{j}} \,.$$

This concept leads to a multiresolution of $L^2(\mathbb{R})$:

$$L^2(\mathbb{R}) = V_{\mathrm{J}} \oplus W_{\mathrm{J}} \oplus W_{\mathrm{J}+1} \oplus W_{\mathrm{J}+2} \oplus \cdots$$

and to a multiresolution, i.e. a decomposition, of the function $f \in L^2(\mathbb{R})$ into orthogonal components

$$f(t) = A_{\mathrm{J}}(t) + \sum_{\mathrm{j} \geq \mathrm{J}} D_{\mathrm{j}}(t) \,,$$

where

$$D_{\mathrm{j}}(t) = \sum_{\mathrm{k} \in \mathbf{Z}} d_{\mathrm{j\,k}}\, \psi_{\mathrm{j\,k}}(t)$$

is the detail on level j, i.e. on scale 2^{-j} and

$$A_J(t) = \sum_{j<J} D_j(t)$$

is the approximation of f in V_J . Figure 3 provides an overview of the decompositon of a signal according to the multiresolution analysis.

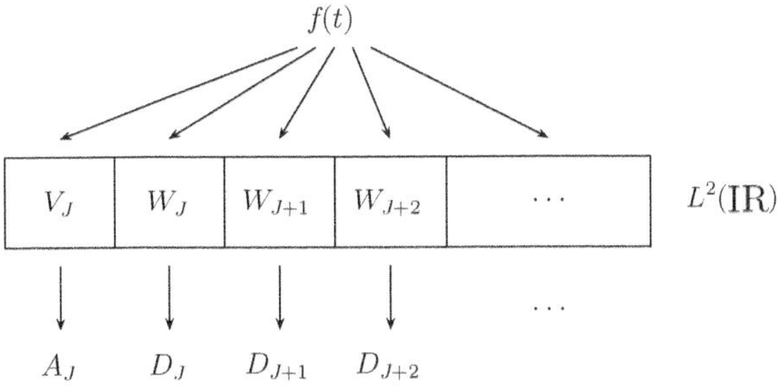

Fig. 3. Decomposition of a function into orthonormal components

Wavelets have several advantages that make them so popular [14]:

(i) they constitute an orthonormal basis,
(ii) they are local in time via translation, and in space via dilation, which, for example, is not true for Fourier transform,
(iii) the coefficients are a measure of the local behaviour, depending on the spatial parameters j and k,
(iv) it is very easy to apply them to functions of more than one variable.

1.2 The Final Model Under the Wavelet Approach

On using a wavelet basis, the following model results

$$f_i(t) = g_i(t) + \varepsilon_i(t) \quad \text{for} \quad i \in \{1, \ldots, n\}$$

$$g_i(t) \in V_N \qquad \text{i.e.} \quad g_i(t) = A_J(t) + \sum_{J \le j < N} D_j(t) \tag{2}$$

$$\varepsilon_i(t) \in V_N^{\perp} \qquad \text{i.e.} \quad \varepsilon_i(t) = \sum_{j \ge N} D_j(t) \, ,$$

where $g_i(t)$ represents the systematic part in the curves observed, i.e. the characteristic feature that should be investigated, and $\varepsilon_i(t)$ stands for the random error. A projection of the curves observed onto this subspace V_N separates the systematic components of the feature observed from the random ones.

The question remaining is, which wavelet basis should be used in (2) and which scale (resolution) should be chosen? Contrary to Fourier transform the wavelet transform is not unique. There is no rule of thumb that prescribes the proper basis for the current problem.

In many practial applications Daubechies wavelets are applied [3, 5]. They have compact support, which is related to computational efficiency [10]. Besides this, they also have some vanishing moments, which improves computational efficiency, since the higher the number of vanishing moments, the more information will be concentrated in a smaller number of wavelet coefficients, and the fine scale wavelet coefficients will be essentially zero where the function is smooth. However, this increases the support of the wavelets, so that a trade-off is necessary [10, 12].

This study uses Daubechies wavelets with two vanishing moments. Different levels of J were also investigated, and it turned out that $J = 5$ led to the lowest classification error in the wavelet model with PCA.

The number of observation knots turns out to play an important role when working with discretized signals ([3, 11] and the references there). Multiresolution analysis requires that the sample size is 2^n for some integer n. In cases where this condition is not satisfied, the problem is often solved by padding the signal with zeros. This procedure usually introduces unnecessary edge effects because of the resulting discontinuity of the signal at the borders. These edge effects are difficult to compensate for. For the data underlying this study zero padding leads to very low classification error rates.

2 Classification Method and Results

The classification of the data was carried out according to the minimum of Mahalanobis distance. As preparation for this, some basic transformations were conducted, such as the log-transformation of the observations and the standardisation of the observation range to an interval starting at zero, this interval being devided into subintervals of length one. As each sample was measured from three positions, the mean of these measurements was calculated and a baseline correction was carried out. Petrological examination of the samples should ensure that the samples consisted only of one type of granite. The signal resulting was projected onto V

$$P : L^2(\mathbb{R}) \to V = span(\{\phi_{-5\,k}, \psi_{-5\,k} \mid k \in \mathbf{Z}\}),$$

so that $f_i = g_i + \varepsilon_i$ and $g_i = A_5^i + D_5^i$. The other details were ignored because different calculations have shown that the classification results improved, when details were not used. This model leads to a reduction in the number of variables by about 95%.

The classification was carried out by use of the coefficients of the principal components, calculated by Principle component analysis (PCA) or PLS estimation. In PCA and PLS, respectively, the number of features that enter the classification is reduced to a further extent. Although this reduction is not overwhelming – 16 scores remain in PLS and in PCA analysis (reduction of about 1.5%!) – it has to be stated that the scores corresponding to low eigenvalues are important in cases when the direction of separation is orthogonal to the first PCs (confer [4]).

Using this wavelet approach, complete classification of the mean measurements was attained. The results of assigning single measurements in terms of the leave-one-out model are summarized in the Tables 2 and 3 below and depend on the additional method of dimension reduction.

Table 2. Results of statistical classification, based on measurement M for the ten samples or particles of each of Granite 1 and Granite 2. Wavelet model with PCA applied to the log-spectra by use of Daubechies wavelets of order two on level five. M_0 denotes the mean measurements, M_1 to M_3 the single measurements, and G_1 and G_2 stand for the two types of granite

	assignment of							
	M_0		M_1		M_2		M_3	
to / from	G_1	G_2	G_1	G_2	G_1	G_2	G_1	G_2
G_1	10	0	9	1	10	0	10	0
G_2	0	10	1	9	1	9	0	10

The three measurements for each sample can be found in the columns M_1 to M_3 of Table 2, showing some misclassifications, whereas the classification based on mean values gives correct assignments. This raises the question, whether the assignment, based on the single measurements M_1 to M_3, and showing misclassification, belongs to the same particle or different particles; unfortunately these are two different particles. But as can be seen in Table 2 the classification error rate for the single measurements is only $\frac{3}{60} = 0.05$, which is very low for curves that are extremely similar as in the case here.

As the dimensionality of the data was only slightly reduced by PCA one could ask, whether it was necessary to use PCA and whether PLS would improve the results. The calculations showed that the error rate was unacceptable in the case when classification was carried out only by use of wavelet coefficients. Therefore, a further data reduction method was applied. After calculation of the PLS estimation of the scores, instead of a PCA, the results improved, but the number of scores was the same as in PCA, i.e. 16 scores

Table 3. Results of statistical classification, based on measurement M for the ten samples or particles of each of Granite 1 and Granite 2. Wavelet model with PLS applied to the log-spectra by use of Daubechies wavelets of order two on level five. M_0 are the mean measurements, M_1 to M_3 the single measurements, and G_1 and G_2 stand for the two types of granite

	assignment of							
	M_0		M_1		M_2		M_3	
from / to	G_1	G_2	G_1	G_2	G_1	G_2	G_1	G_2
G_1	10	0	10	0	10	0	10	0
G_2	0	10	1	9	1	9	0	10

were used in further classification. Table 3 shows that only two single curves were misclassified after application of PLS for further dimension reduction. These two curves belong to different samples.

PLS is often applied in chemometrics to reduce the dimension of spectral data for classification. The following Table 4 gives an impression of the classification error, which arises when PLS is used to select the original log-transformed spectral lines. This means that PLS is used for identifying the relevant observation knots. As can be seen from Table 4, there is even

Table 4. Results of statistical classification, based on measurement M for the ten samples or particles of each of Granite 1 and Granite 2. PLS applied to the log-spectra. M_0 denotes the mean measurements, M_1 to M_3 the single measurements, and G_1 and G_2 stand for the two types of granite

	assignment of							
	M_0		M_1		M_2		M_3	
from / to	G_1	G_2	G_1	G_2	G_1	G_2	G_1	G_2
G_1	9	1	7	3	9	1	9	1
G_2	1	9	1	9	0	10	2	8

classification error in the assignment of the mean measurements. As three measurements of each sample were classified, the question on the number of misclassified samples turns up again. In the case of the single measurements, 8 curves but only 5 samples were incorrectly assigned.

3 Summary

After a proper use of statistical spectra approximation it is possible to separate variants of granite, even if they are very similar in terms of their reflectivity of mid-infrared light. As reflectivity can vary due to the measurement positions, classification accuracy is improved by use of several measurements from different positions of the sample. For such mean measurements, a reliable prediction of the class membership can easily be derived by use of wavelets. A further reduction of the number of classification features using PCA or PLS is necessary. The use of PLS for estimation of the scores appeared to be superior to the traditional PCA in the sense that the variability in measurements have led to less of a deterioration in the classification results. The wavelet model is superior to the PLS based reduction in the dimensionality of the original data.

References

1. EUREKA (2004), PETROSCOPE – An Optical Analyser for Construction Aggregates and Rocks, Project no. 2569, Announced 28 June 2001, EUREKA, Brussels
2. Hunt, G.R. (1987) Spectroscopic properties of rocks and minerals, in: Carmichael, R. S., (ed.), Handbook of Physical Properties of Rocks, CRC Press, Boca Raton, 295–385
3. Jetter, K., Depczynski, U., Molt, K. and Niemöller, A. (2000) Principles and applications of wavelet transformations to chemometrics, Analytica Chemica Acta **420**, 169–180
4. Jolliffe, I.T. (1986) Principle Component Analysis, Springer Verlag, Berlin
5. Leung, A.K., Chau, F. and Gao, J. and (1998) A review of wavelet transform techniques in chemical analysis: 1989–1997, Chemometrics and Intelligent Laboratory Systems, **43**, 165–184
6. Müller, F. (1987) Gesteinskunde, Ebner Verlag, Ulm
7. Ramsay, J.O. and Silverman, B.W. (1997) Functional Data Analysis, Springer Series in Statistics, Springer, New York
8. Ramsay, J.O., Silverman, B.W. (2001) Applied Functional Data Analysis, Springer, New York
9. Schmidt, W. (1994) Optische Spektroskopie, Eine Einführung für Naturwissenschaftler und Techniker, VCH, Weinheim
10. Teppola, P. and Minkkinen, P. (2000) Wavelet-PLS regression models for both exploratory data analysis and process monitoring, Journal of Chemometrics, **14**, 383–399
11. Trygg, J. and Wold, S. (1998) PLS regression on wavelet compressed NIR spectra, Chemometrics and Intelligent Laboratory Systems, **42**, 209–220
12. Vannucci, M., Brown, P. J. and Fearn, T. (2003) A decision theoretical approach to wavelet regression on curves with a high number of regressors, Journal of Statistical Planning and Inference **112**, 195–212
13. Vidakovic, B. (1999) Statistical Modeling by Wavelets, Wiley, New York
14. Vidakovic, B. and Müller, P. (1999) An Introduction to Wavelets, in Müller, P., Vidakovic, B., (eds.), Bayesian Inference in Wavelet-Based Models, Springer, New York
15. VTT Electronics (2001) Petroscope Prestudy. Confidential Report to Petromodel Ltd, VTT, Oulu, Finland

Part II

Geostatistical Applications

Simulating the Effects of Rural Development Policies on Land Use: Evidence from Spatially Explicit Modeling in the Central Highlands of Vietnam

Daniel Müller[1] and Darla K. Munroe[2]

[1] Leibniz Institute of Agricultural Development in Central and Eastern Europe, Halle (Saale), Germany
mueller@iamo.de

[2] Department of Geography, The Ohio State University, Columbus, OH, USA
munroe.9@osu.edu

1 Introduction

Land cover, the spectral characteristics of the earth' surface, and land use, the operational employment on that land, are closely related. However, there is also a clear distinction between land use and land cover. While land cover refers to the biophysical earth surface, land use is shaped by human, socioeconomic and political influences on the land [7]. In essence, 'land use links land cover to the human activities that transform the landscape' [15]. In most practical applications the analysis of satellite images are used to infer land use from land cover.

Land use is a common phenomenon associated with population growth, market development, technical and institutional innovation, and related rural development policy. This paper attempts to assess the impact of policy, technology, socioeconomic, and geophysical conditions on land use in the last decade and combines data from a village-level survey with remote sensing data derived from Landsat images. Our objective is to analyze the influence of these explanatory variables on land use using a reduced-form, spatially explicit multinomial logit model. Simulations are then carried out to assess the effects of three policy scenarios of rural development on land use. An empirical application is presented for two districts of Dak Lak province in the Central Highlands of Vietnam. Dak Lak exhibits an interesting case in the study of land use dynamics with its abundant forest resources, ethnic diversity, high immigration rates and dynamic agricultural and socioeconomic development. In particular, the last decade was characterized by rapid, labor- and capital-intensive growth in the agricultural sector.

2 Methodology and Data

Land use is approximated by visual interpretation of a Landsat Enhanced Thematic Mapper (ETM) image from 2000. The resulting land-use map is categorized into three classes: (1) mixed agricultural land (cash and food crops) including cultivated upland plots; (2) paddy fields, both with one and two crops a year; (3) all non-agricultural land comprising forests of different quality and small areas of mixed grassland as well as fallowed upland plots. Results of earlier studies showed only modest land-use changes in the last decade in absolute and relative terms. Forest cover increased slightly mainly due to agricultural intensification, better market access and government policies on forest protection [10].[3]

We explain land use as a function of four types of regressors. First, we use geophysical indicators such as the level and variance of rainfall derived from interpolations of point measurements, surfaces of soil suitability for mixed agriculture and paddy rice, the altitude and the slope of land. Second, we employ historical variables that describe socioeconomic characteristics of the villages. Here, we capture endogenous, and therefore lagged, population as a time-variant variable to indicate the state at the beginning of the period. A continuous population surface was generated from village recall data for 1990 by interpolating the point coverages of village locations with inverse distance weighting [5]. Cultural influences on land use are captured by a dummy for the ethnic composition of the villages. Third, exogenous policy variables to describe government investments and macro-level policies are included in the model. These include a quantification for the spatial placement of policy-induced investments in road and market infrastructures and the introduction of agricultural technologies, proxied by the increase in irrigation per village and the year of introduction of a compound fertilizer containing nitrogen, phosphorus and potassium (NPK) as a yield-increasing input. To capture market access we used a road network layer passable during the whole year and existing since the time of French colonial rule in the first part of the 19th century. Therefore, we assume the road layer to be exogenous to present-day land use. In addition, areas delineated as New Economic Zones (NEZ)[4] were incorporated as a measure of the length since the inauguration of government-controlled immigration and associated investments. To incorporate the patchiness of land use we include one indicator for landscape fragmentation, measured by the ratio of perimeter to area of a particular landscape patch. The lagged value of this variable is a proxy the spatial structure of individual landscape elements. Due to possible scale economies in a

[3] Additional analysis and discussion of these issues are to be found in Müller [8] and Müller and Munroe [9].

[4] A resettlement scheme undertaken by the government after the end of the war in 1975, where people from densely populated areas such as the Red River Delta and the Mekong Delta were moved to less-densely settled areas for agricultural production.

situation of agricultural intensification and increasing market orientation, we expect that pixels under intense agricultural production will tend to be more spatially concentrated in areas close to market centers and in favorable natural conditions. Agricultural land uses are, therefore, anticipated to show more homogeneous patterns.

We will explore the causal relationships between these variables in a spatially explicit framework, quantify their respective direction and magnitude, and simulate the effect of rural development policies. Detailed descriptive statistics of the independent variables are reported in Table 1.

Table 1. Descriptive statistics for independent variables

variable	mean	std.dev.	min.	max.	N
mixed agriculture	0.24	0.43	0	1	22,321
paddy	0.08	0.27	0	1	22,321
non-agricultural land	0.68	0.47	0	1	22,321
slope (degrees)	10.18	9.02	0	42	22,321
spatial lag slope (degrees)	21.01	16.74	0	77.9	22,321
elevation (100 m)	5.88	1.67	4.2	13.1	22,321
soil suitability for paddy (0/1)	0.14	0.35	0	1	22,321
soil suitability for mixed agriculture (0/1)	0.04	0.19	0	1	22,321
rainfall sum (100 mm), 1992–1999	16.41	1.78	9.9	18.6	22,307
rainfall variance (100 mm), 1992–1999	27.62	7.07	16.7	41.5	22,307
population, 1990	65.09	26.67	0.2	286	22,294
ethnic minority village (0/1)	0.62	0.49	0	1	22,321
distance to all-year road (km)	5.86	5.36	0	23.1	22,321
years since establishment of NEZ	9.26	8.55	0	24	22,321
years since introduction of NPK	4.26	3.7	0	20	22,321
increase in irrigated area (ha)	115.32	202.88	0	938	22,321
protected area Nam Ca (0/1)	0.08	0.27	0	1	22,321
landscape fragmentation, 1992	843.66	280.3	16.6	1,102.6	22,304

Source: Primary data on village level collected in village survey; secondary data on geophysical and agroecological variables were provided by the Mekong River Commission (Digital Elevation Model) and the Department for Agriculture and Rural Development, Dak Lak (Digital Soil Map and protected areas); rainfall data from own interpolation of data from nine meteorological stations, classification of soil suitability dummies from expert opinion.

3 Spatial Sampling

Ideally, to integrate spatially explicit data derived from geographical information systems (GIS) and remote sensing (RS) techniques with village survey data, the scale of the analysis should match the agricultural plots as the unit of decision-making. Yet, in Vietnam as in most developing countries, plot maps and village boundaries are not available [12]. This renders spatial modeling a time-consuming and costly task due to the necessary delineation of the spatial extent of plots or villages, e.g. using Global Positioning Systems (GPS). To demarcate the spatial base unit for the integration of socioeconomic variables, the geographic positions of all villages were recorded using GPS and point coverages created. Village boundaries for all surveyed villages were then approximated by applying a cost-distance algorithm to delineate a set of explicitly defined 'accessibility catchments', generated around each village location and based on estimated transport costs [4]. Spatial accessibility is similar to Euclidean distance functions, but instead of calculating the actual distance from one point to another, the shortest cost distance (or accumulated transportation cost) from each cell to the nearest source cells is determined (Fig. 1).[5] The resulting catchment polygons were then used as a base unit for village-level data in sub-sequent analysis. Hence, survey data – apart from population – takes the value of the interviewed village for each point, which has lower transportation costs to the geographic location of that village than to any other village location.

The units of analysis are square pixels of 50 by 50 m, i.e. 0.25 ha. To focus on changes at the forest margins influenced by human interventions, we restrict the analysis to those pixels that have a cost of access below the mean for the

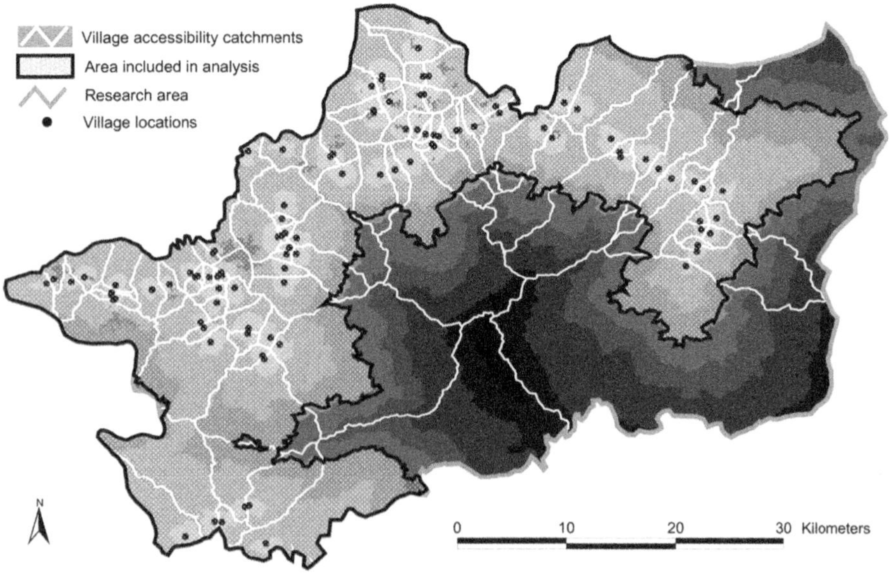

Source: authors

Fig. 1. Transportation cost surface with spatial sample and approximated village borders [9]

[5] For an example of spatial data integration using purely Euclidean distance measures, see [10].

transportation cost surface (Fig. 1). In that way, we include nearly all the agricultural area in 2000 and eliminate remote and high mountainous areas covered mostly with thick primary forest, which are outside measurable human influence.

At present, there are no models and test statistics available to account for substantive spatial interaction in a qualitative dependent variable framework [2]. To compensate for potential spatial dependence in the dependent variables, Besag's coding scheme was used [3], also employed by [14] and [11] in similar studies. The regular spatial sample was drawn by selecting every 5th cell in the X and Y directions so that no selected cells are physical neighbors. The sampling procedures allow us to apply standard estimation techniques [1] and resulted in a dataset of 22,300 observations used for subsequent econometric modeling. In addition, we include slope as a spatially lagged variable [11, 13, 14]. These techniques help to reduce spatial autocorrelation although they may not totally eliminate it [6].

4 Models and Results

4.1 Econometric Estimation

To explore relationships between exogenous and predetermined variables and the land cover categories as left-hand side variables, a multinomial logit specification (MNL) was applied. MNL models estimate the direction and intensity of the explanatory variables on the categorical dependent variable by predicting a probability outcome associated with each category of the dependent variable. MNL is based on the assumption that the probabilities are independent of other outcomes.

Assuming that pixels are independent across villages, but not necessarily within, we cluster all pixels based on approximated village areas. This affects the estimated standard errors and the variance-covariance matrix of the estimators, but not the estimated coefficients [16]. Further, we employ the Huber and White sandwich estimator to obtain robust variance estimates.

The sampling procedure outlined in Sect. 3 resulted in 22,300 observations, which we use to estimate the coefficients of the MNL. These coefficients are then used to predict outcomes for the entire dataset (558,000 observations) in order to generate continuous prediction and simulation maps.

4.2 Empirical Results

The MNL has three land cover classes as categorical, unordered dependent variables. To control for potential endogeneity problems, only lagged values for time-variant independent variables such as population growth and road access are considered in the empirical applications. In addition, all variables were tested for multicollinearity. We assess the assumption of independence

Table 2. Multinomial logit results (non-agricultural land use as base category)

	Mixed agriculture	paddy
slope	**–0.088**	**–0.207**
	(6.97)***	(4.28)***
spatial lag slope	**–0.046**	**–0.072**
	(7.51)***	(4.74)***
elevation	**–2.621**	**–6.695**
	(4.56)***	(6.73)***
soil suitability for paddy	**1.306**	**2.543**
	(5.52)***	(8.97)***
soil suitability for mixed agriculture	**1.5**	**1.047**
	(4.51)***	(2.74)***
rainfall sum, 1992–1999	**0.504**	**1.293**
	(1.90)*	(1.97)**
rainfall variance, 1992–1999	**–0.116**	**–0.265**
	(2.86)***	(3.85)***
population, 1990	0.007	0.013
	–0.82	–0.97
ethnic minority village	–0.123	–0.433
	–0.32	–1.13
distance to all-year road	–0.001	–0.004
	–0.02	–0.04
years since establishment of NEZ	0.041	–0.022
	–1.6	–0.63
years since introduction of NPK	**0.062**	**0.128**
	(1.65)*	(2.95)***
increase in irrigated area	0	**0.001**
	–0.78	(2.25)**
protected area Nam Ca	**–2.708**	**–2.74**
	(7.71)***	(4.85)***
landscape fragmentation, 1992	**–0.017**	**–0.02**
	(4.07)***	(3.02)***
constant	6.488	11.168
	–1.61	–0.95
observations	22,255	22,255

Source: own calculations.

of irrelevant alternatives using both the Hausman and the Small-Hsiao test and can accept the null hypothesis that outcomes are independent of other alternatives. Model results are reported as raw coefficients in Table 2 for non-agricultural land as the comparison group. Overall predictive power is 88%, measured as the locations predicted correctly. Equivalent to [13] we found that wrong predictions frequently lie on the border between land use classes, which is likely to be related to spatial errors in the source data and artifacts inherent in our technique of data integration.

Coefficients for the geophysical variables are mostly significant at the 1% level and show the expected signs with high predictive power. Agriculture is more likely at lower altitudes, flatter land, and more suitable soils. More amount and less variance of rainfall increases the likelihood of paddy compared to the other two categories. Surprisingly, access to all-year roads did not have a significant effect on the probability of a certain land use class. We assume that this is due to the relatively large areas under agricultural uses far away from the all-year road network. Earlier introduction of mineral fertilizer and more irrigated area increases the likelihood that pixels are under paddy production. Lagged population does not seem to have an influence on the amount of area cultivated as it was probably outweighed by effects from agricultural intensification. In addition, the majority of migrants settled in areas with high proportions of land suitable for paddy cultivation. Therefore, lagged population has little influence on the amount of land used for cultivation. The dummy on ethnic composition is significant at the 5% level and has a strong negative effect on the probability of paddy land. A more fragmented landscape in an earlier period decreases the likelihood of present agricultural uses and a significant amount of fragemented agricultural plots regenerated into non-agricultural uses. Finally, forest protection has a strong effect on the likelihood to observe both mixed agriculture and paddy land.

4.3 Policy Simulations

Changes in socioeconomic variables can be simulated by changing the values of the respective explanatory variables. Applying the estimated coefficients from the base estimation for the entire data set, new predictions and probabilities for land use can then be generated with simulated values of explanatory right-hand-side variables. Comparing simulated predictions of land use to the base predictions yields an approximation of the effects of varying levels of explanatory variables on land use [9, 13]. One main advantage of the spatially explicit estimation is the possibility to assess locational changes and to identify potential hot spots of land-use change following certain policy interventions.

We distinguish three policy scenarios for rural development interventions (see Table 3). The first scenario assumes an earlier introduction of NPK fertilizers. Governments can influence such technology adoption by increasing

Table 3. Simulated policy scenarios

description		proxy
1. Earlier introduction of fertilizer	→	introduction of NPK 5 years earlier
2. Forest protection	→	(a) protection of existing primary forest
	→	(b) protection on slopes > 15 degrees
3. Earlier introduction of fertilizer and forest protection	→	scenarios 2 and 3 combined

Source: authors.

extensions services and by supporting fertilizer application through price incentives. The 'forest protection scenario' follows a government guideline that discourages agricultural production above slopes of 15 degrees. We assume here that this sloping land and, in addition, all primary forest with closed crown cover are added to already existing protected areas. The last policy option combines 'earlier introduction of NPK' with the 'forest protection scenario' to assess the effect of this predominant policy strategy in Vietnam. Because of the insignificant coefficients, we did not simulate the upgrading of seasonal roads to year-round access as it would not result in considerable land-use changes.

A comparison of simulated changes with the baseline predictions is expected to shed light on the impact of the rural development policy scenarios on land use. The spatially explicit framework allows for an assessment of both the magnitude as well as the location and spatial arrangement of the simulated changes. This will enable the identification of areas, which are likely to change into other forms of land use under a certain policy setting [9].

5 Simulation Results

The scenario of 'earlier introduction of fertilizer' results in a reduction in non-agricultural land uses by 14.4 km^2 and an intensification of mixed agricultural land into paddy of 22 km^2. The observation of the resulting prediction map reveals that most of the land switching from non-agriculture was bare soil and grass land. However, the spatial fragmentation of the simulation map as measured by the number of patches increases by 6%. The scenario of 'forest protection' decreases agricultural land use by 4.8 km^2. An increasing number of patches leads to more scattered spatial arrangements, which is less desirable from an ecological viewpoint. The result suggests that forest protection strategies ought to be combined with ecological valuations that explicitly take into account the value of contiguous protection areas conserving precious biodiversity.

Due to space limitations we concentrate the following discussion on the results of the third policy scenario. Significant changes in the spatial patterns of land use result from a combination of policies to boost the introduction of yield-increasing technologies combined with forest protection (Table 4 and Fig. 2). It induces a reduction in area under mixed agriculture by 26.9 km^2. The intensification effect of an earlier introduction of NPK fertilizer increases the probability for paddy cultivation while mixed agricultural land retreats on marginal land following the simulated increase of forest protection.

Figure 2 shows the spatial changes resulting from the policy interventions (see also [8]). On flat land and in suitable areas agriculture is intensified as can be observed in the northwestern part of the analyzed area. Agricultural

Table 4. Earlier introduction of NPK combined with forest protection

base prediction	simulated prediction: mixed agriculture	paddy	non-agricultural land	total
mixed agriculture	347.5	21.9	5.0	**374.3**
paddy	0.0	88.0	0.0	**88.0**
non-agricultural land	12.7	0.4	940.6	**953.8**
total	**360.2**	**110.3**	**945.6**	**1,416.1**

Source: own calculations; numbers reported in km^2.

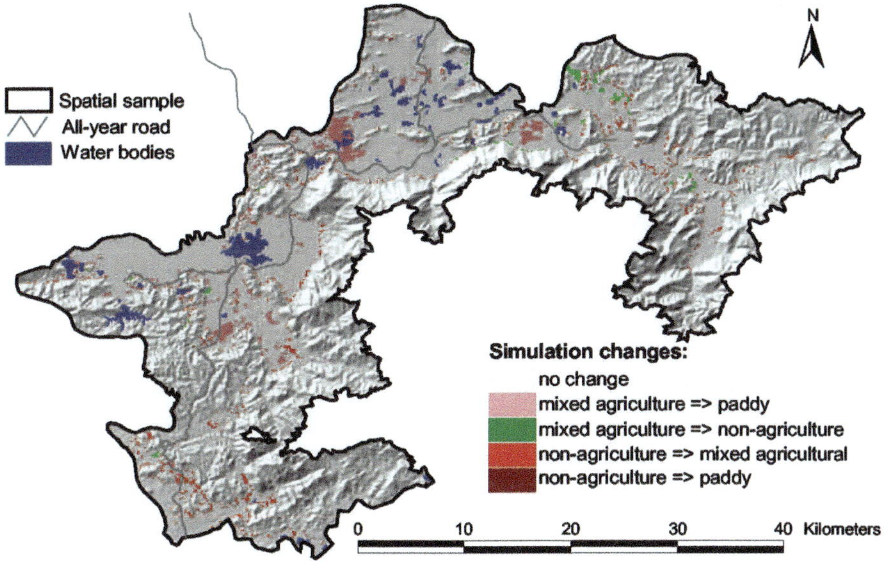

Fig. 2. Prediction maps of policy scenarios on land use compared to base prediction

production is left abandoned in sloping areas close to the lowlands as in the south and northeast. The identification of such potential hot-spots calls for more in-depth field visits to monitor resulting land-use outcomes.

6 Discussion and Policy Implications

Agricultural land uses are concentrated on lower altitudes, flatter land, and on better soils. Forest protection has a strong effect on the likelihood to observe both mixed agriculture and paddy land. An earlier introduction of yield-increasing fertilizers increases and higher investments in irrigation increase the probability of paddy land, while in ethnic villages paddy land is less likely. The changes in land use over the last decade show that agricultural intensification and improved market access made farming more profitable and labor-intensive. This led to higher outputs that compensated for the decline in the availability of land per capita given the high population growth rates. Shifting cultivation as the traditional land-intensive farming system practiced by the indigenous population in the research area almost entirely disappeared during the last decade [8, 10].

The spatially explicit simulations demonstrate possible land-use changes resulting from rural policy interventions. Government investment in agricultural intensification, proxied by an earlier introduction of NPK fertilizers, has a strong positive influence on the area under paddy rice, by far the most important food crop for farmers in the research area. The forest protection scenario induced higher pressure on agricultural land uses within the simulated additional areas under forest protection and resulted in the abandonment of more marginal lands. The main advantage we see in this methodology is the facilitation of a spatial assessment of land-use changes. In that way, the identification of hot spots, possibly requiring additional conservation efforts, and areas with untapped agricultural potentials becomes possible.

The two districts in our study were purposively selected and can, therefore, not be generalized for the whole of Dak Lak province. Nonetheless, the policy implications of this study call for a renewed emphasis on rural and agricultural development that can address the subsistence and income needs through technological progress and agricultural intensification. If combined with forest protection, especially in locations with crucial watershed and biodiversity functions, agricultural intensification might safeguard locally and globally valuable environmental services. With respect to implications for further research, we conclude that problems for spatially explicit modeling and spatial statistics are found more frequently in the combination of data on natural resources with socioeconomic information at an acceptable scale and under reasonable assumptions and simplifications. Our attempt to combine land use data with accessibility catchments at the village level addresses this issue. However, this aggregation masks decision-making processes at the farm and plot levels. More disaggregated data is needed to consider the effects of rural development policies on cropping patterns and micro-level changes in a spatially explicit framework.

To obtain better spatial information for socioeconomic indicators, censuses would yield more variation in explanatory indicators, improve the strength of the estimations, and facilitate more sophisticated spatially explicit policy simulations. Combined with landscape metrics, more disaggregated data can yield additional insights into the spatial composition and arrangement of potential land-use changes resulting from investments in rural development. In that way, spatially explicit econometric modeling can enhance the geographical targeting of rural development interventions and facilitate the assessment of the magnitude and location of their economic and environmental consequences.

Acknowledgments

We are grateful to Manfred Zeller and Regina Birner for useful comments and suggestions on previous versions of this paper. The research was funded by the German Ministry of Economic Development and Cooperation (BMZ) under the Tropical Ecology Support Programme (TÖB) of the German Agency for Technical Cooperation (GTZ). The writing of this chapter was supported by the German Research Foundation (DFG) under the Emmy Noether-Programme.

References

1. Luc Anselin (2001) Spatial econometrics. In Badi Baltagi, editor, *Companion to Econometrics*, pages 310–330. Basil Blackwell, Oxford.
2. Luc Anselin (2001) Spatial effects in econometric practice in environmental and resource economics. *American Journal of Agricultural Economics*, 83(3):705–710.
3. Julian Besag (1974) Spatial interaction and the statistical analysis of lattice systems. *Journal of the Royal Statistical Society*, Series B 36:192–236.
4. David Bigman and Hippolyte Fofack (2000) *Geographical targeting for poverty alleviation: Methodology and applications.* World Bank Regional and Sectoral Studies Series. World Bank, Washington, DC.
5. Ian Bracken (1994) A surface model approach to the representation of population-related social indicators. In Stewart A. Fotheringham and Peter Rogerson, editors, *Spatial Analysis and GIS*, pages 247–259. Taylor & Francis, London.
6. Alessandro De Pinto and Gerald C. Nelson (2002) *Correcting for Spatial Effects in Limited Dependent Variable Regression: Assessing the Value of "Ad-Hoc" Techniques.* Paper prepared for the American Agricultural Economics Association annual meeting, Long Beach, California.
7. Helmut J. Geist and Eric F. Lambin (2002) Proximate causes and underlying driving forces of tropical deforestation. *BioScience*, 52(2):143–150.
8. Daniel Müller (2003) Land-use Change in the Central Highlands of Vietnam: A spatial econometric model combining satellite imagery and village survey data. Doctoral dissertation, Georg-August University Göttingen, Göttingen.
9. Daniel Müller and Darla K. Munroe (2005) Tradeoffs between rural development policies and forest protection: Spatially explicit modeling in the Central Highlands of Vietnam. *Land Economics*, 81(3):412–425.
10. Daniel Müller and Manfred Zeller (2002) Land use dynamics in the central highlands of vietnam: A spatial model combining village survey data and satellite imagery interpretation. *Agricultural Economics*, 27(3):333–354.
11. Darla K. Munroe, Jane Southworth, and Catherine M. Tucker (2002) The dynamics of land-cover change in western honduras: Exploring spatial and temporal complexity. *Agricultural Economics*, 27(3):355–369.

12. Gerald C. Nelson and Jacqueline Geoghegan (2002) Deforestation and land use change: sparse data environments. *Agricultural Economics*, 27(3):201–216.
13. Gerald C. Nelson, Virginia Harris, and Steven W. Stone (2002) Deforestation, land use, and property rights: Empirical evidence from darien, panama. *Land Economics*, 77(2):187–205.
14. Gerald C. Nelson and Daniel Hellerstein (1997) Do roads cause deforestation? using satellite images in econometric analysis of land use. *American Journal of Agricultural Economics*, 79:80–88.
15. Policy Division Committee on Global Change Research NRC (National Research Council), Board on Sustainable Development (1999) *Global Environmental Change: Research Pathways for the Next Decade.* National Academy Press, Washington, DC.
16. StataCorp (2003) *Stata Statistical Software: Release 8.0.* Stata Corporation, College Station, Texas.

Kriged Road-Traffic Maps

Hans Braxmeier[1], Volker Schmidt[2], and Evgeny Spodarev[3]

[1] Department of Applied Information Processing, University of Ulm, Ulm, Germany
hans.braxmeier@uni-ulm.de
[2] Department of Stochastics, University of Ulm, Ulm, Germany
volker.schmidt@uni-ulm.de
[3] Department of Stochastics, University of Ulm, Ulm, Germany
evgeny.spodarev@uni-ulm.de

1 Introduction

A common difficult problem of large cities with heavy traffic is the prediction of traffic jams. In this paper, a first step towards mathematical traffic forecasting, namely the spatial reconstruction of the present traffic state from pointwise measurements is briefly described. For details, we refer to [1], where models of stochastic geometry and geostatistics are used to spatially represent the traffic state by means of velocity maps. A corresponding Java software that implements efficient algorithms of spatial extrapolation is developed; see [5].

To illustrate our extrapolation method, we use real traffic data originating from downtown Berlin. It was provided to us by the Institute of Transport Research of the German Aerospace Center (DLR). Approximately 300 test vehicles (taxis) equipped with GPS sensors transmit their geographic coordinates and velocities to a central station within regular time intervals from 30 s up to 6 min; see Fig. 2. Thus, a large data base of more than 13 million positions was formed since April 2001; see Fig. 1.

Fig. 1. Observed positions of test vehicles in downtown Berlin

In the first stage of our research, only a smaller data set (taxi positions on all working days from 30.09.2001 till 19.02.2002, 5.00–5.30 pm, moving taxis only) was considered. Furthermore, the observation window was reduced to downtown Berlin to avoid inhomogeneities in the taxi positions.

The main idea of the extrapolation technique described in Sects. 2 and 3 below is to interpret the velocities of all vehicles at given time t as a realization of a spatial random field $V(t) = \{V(t,u)\}$ where $V(t,u)$ is a traffic velocity vector at location $u \in \mathbb{R}$ and time instant $t \geq 0$. The goal is to analyze the spatial structure of these random fields of velocities in order to describe the geometry of traffic jams. Since $V(t,u)$ can be measured just pointwise at some observation points $u_1, \ldots, u_n$, a spatial extrapolation of the observed data is necessary. Notice that the velocities strongly depend on the location and the direction of movement, e.g. the speed limits and consequently the mean velocities are higher on highways than in downtown streets.

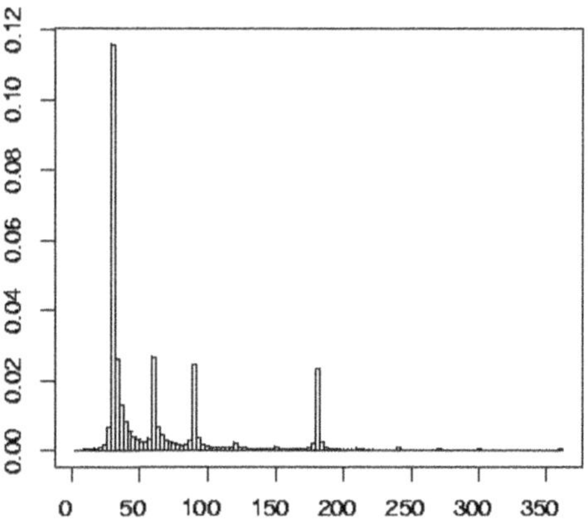

Fig. 2. Histogram of time intervals between consecutive GPS signals (in s)

The classical extrapolation methods of geostatistics (see e.g. [6]) either make no use of additional directional information or provide measurements $V(t, u + u_i)$ and $V(t, u - u_i)$ with equal weights. Both these features are not adequate to the setting mentioned above. Thus, the standard extrapolation methods had to be adapted to our specific problem. In Sect. 3, an ordinary kriging with moving neighborhood is described that allows to extrapolate directed velocity fields. First, the original data set should be split into four subsets which are directionally homogeneous. A data unit $(u, V(t,u))$ belongs to the data set i $(i = 1, \ldots, 4)$ if the polar angle of the vector $V(t,u)$ lies within the directional sector $S_i = \{\alpha \in [0, 2\pi) : (i-1)\pi/2 \leq \alpha < i\pi/2\}$. By convention, the zero polar angle corresponds to the eastward direction on the city map. The above data sets should be extrapolated separately for each directional sector. This yields four velocity maps corresponding to the four sectors of directions.

In what follows, the data from a given time interval, i.e. $[5.00, 5.30]$ pm, will be taken for extrapolation. Keeping this in mind, we shall omit the time parameter t in further notation.

The extrapolation method described in Sects. 2 and 3 has been implemented in Java, where a software library has been developed comprising the estimation and fitting of variograms as well as the ordinary kriging with moving neighborhood; see [5]. As far as it is known to the authors, this is the first complete implementation of such kriging methods in Java. Much attention was paid to the efficient implementation of fast algorithms. In contrast to classical geostatistics operating with relatively small data sets, this efficiency is of great importance for larger data sets with more than $10,000$ entries; see [1] for details.

In Sect. 4, a numerical example is discussed which shows how the developed extrapolation technique can be applied to directional traffic data. Some structural features of the resulting velocity maps (see Figs. 5 and 6) are also discussed. In Sect. 5, this is combined with a statistical space-time analysis of polygonal road-traffic trajectories which have been extracted from the original traffic data. For example, it turns out that the distribution of the number of segments in these traffic trajectories can be fitted quite well by a geometric distribution. The directional distribution of the segments reflects the anisotropy of the street system of downtown Berlin, where the distribution of segment lengths is demonstrably non-normal. Furthermore, the distributions of velocity residuals, i.e. the deviations from their means, show interesting skewness properties which depend on the considered classes of low, medium, and high mean velocities, respectively. A short outlook to simulation and prediction of future traffic states is given in Sect. 6.

2 Random Fields

To model traffic maps, non-stationary random fields composed of a deterministic drift and an intrinsically stationary random field of order two (residual) are used. See e.g. the monographs [4] and [6] for details.

Let $X = \{X(u), u \in \mathbb{R}^2\}$ be a non-stationary random field with finite second moment $EX^2(u) < \infty$, $u \in \mathbb{R}^2$. Then, $X(u)$ can be decomposed into a sum $X(u) = m(u) + Y(u)$, where $m(u) = EX(u)$ is the mean field (*drift*) and $Y(u) = X(u) - m(u)$ is the deviation field from the mean or *residual*. Assume that $\{Y(u)\}$ is intrinsically stationary of order two. Denote by

$$\gamma(h) = \frac{1}{2}E[(Y(u) - Y(u+h))^2] \qquad (1)$$

its variogram function. In practice, the field X can be observed in a compact (mostly rectangular) window $W \subset \mathbb{R}^2$. Let $x(u_1), \ldots, x(u_n)$ be a sample of observed values of X, $u_i \in W$ for all i. The extrapolation method described in Sect. 3 yields an "optimal" estimator $\widehat{X}(u)$ of the value of $X(u)$ for any $u \in W$ based on the sample variables $X(u_1), \ldots, X(u_n)$.

3 Kriging Based on Residuals

Among the variety of extrapolation techniques for non-stationary random fields, our approach is similar to the so-called *kriging based on residuals*; see [4], p. 190. First of all, an estimator $\widehat{m}(u)$ for the drift $m(u)$ has to be constructed. Then, the deviation field $Y^* = \{Y^*(u),\, u \in \mathbb{R}^2\}$ defined by

$$Y^*(u) = X(u) - \widehat{m}(u) \tag{2}$$

is formed and its kriging estimator $\widehat{Y}^*(u)$ is computed. Finally, the estimator $\widehat{X}(u)$ is given by

$$\widehat{X}(u) = \widehat{m}(u) + \widehat{Y}^*(u)\,. \tag{3}$$

If we suppose that the drift is known, i.e. $\widehat{m}(u) = m(u)$ for all u, then we have exact values of the deviation field $Y(u_1), \ldots, Y(u_n)$ since in this case

$$Y^*(u) = Y(u) = X(u) - m(u)\,.$$

Let $\{y(u_i) = x(u_i) - m(u_i), \quad i = 1, \ldots, n\}$ be a realization of the sample variables $Y(u_1), \ldots, Y(u_n)$. The extrapolation of $Y(u)$ can be performed by *ordinary kriging* making use of the variogram $\gamma(h)$; see [4, 6].

3.1 The Kriging Estimator

A simpler version of the following *ordinary kriging with moving neighborhood* can be found in [3], pp. 201–210 and [6], pp. 101–102. Consider the usual indicator function

$$\mathbf{1}\{x \in B\} = \begin{cases} 1 & \text{if } x \in B, \\ 0 & \text{otherwise.} \end{cases}$$

Introduce the estimator $\widehat{Y}(u)$ of $Y(u)$ for $u \in W$ as a linear combination of the sample variables $Y(u_i)$ with unknown weights $\lambda_i = \lambda_i(u)$ by

$$\widehat{Y}(u) = \sum_{i=1}^{n} \lambda_i Y(u_i) \mathbf{1}\{u_i \in A(u)\}\,. \tag{4}$$

The estimation involves only those sample random variables $Y(u_i)$ that are positioned in the "neighborhood" $A(u)$ of u, i.e. if $u_i \in A(u)$. Being an arbitrary set, this moving neighborhood $A(u)$ contains a priori information about the geometric dependence structure of the random field Y. For instance, it could be designed to model the formation of traffic jams; see Sect. 4.

Unbiasedness of the estimator introduced in (4) and minimizing its variance lead to the following conditions on the weights λ_i. For all $i = 1, \ldots, n$ with $u_i \in A(u)$ it holds

$$\sum_{j=1}^{n} \lambda_j \gamma(u_j - u_i) \mathbf{1}\{u_j \in A(u)\} + \mu = \gamma(u - u_i) \,, \tag{5}$$

$$\sum_{j=1}^{n} \lambda_j \mathbf{1}\{u_j \in A(u)\} = 1 \,.$$

To solve this system of equations, the knowledge of the variogram function $\gamma(h)$ is required. However, in most practical cases $\gamma(h)$ is unknown and has to be estimated from the data $y(u_1), \ldots, y(u_n)$.

3.2 Variograms

In this paper, the most simple and popular variogram estimator of Matheron is used (cf. [3, 6]). It is defined by

$$\widehat{\gamma}(h) = \frac{1}{2N(h)} \sum_{i,j:u_i - u_j \approx h} (Y(u_i) - Y(u_j))^2 \tag{6}$$

where $u_i - u_j \approx h$ means that $u_i - u_j$ belongs to a certain neighborhood $U(h)$ of vector h and $N(h)$ denotes the number of such pairs (u_i, u_j) for $i, j = 1, \ldots, n$.

As shown in Fig. 4, the traffic data lead to empirical variograms that are clearly zonally anisotropic. Below, we consider zonally anisotropic variogram models constructed from isotropic ones (cf. [4, 6]). Put

$$\gamma(h) = \gamma_1(h) + \gamma_2(h) \,, \tag{7}$$

where $\gamma_1(h)$ is an exponential isotropic variogram model with nugget effect $a_1 > 0$, sill b_1 and range c_1. The second term

$$\gamma_2(h) = b_2 \left(1 - e^{-\sqrt{h^\top C h}/c_2}\right) \tag{8}$$

is a geometrically anisotropic exponential variogram model with sill $b_2 > 0$ and a further parameter $c_2 > 0$. For a vector $h = (h_1, h_2)$, the quadratic form

$$h^\top C h = \lambda_2 h_1^2 + \lambda_1 h_2^2 + (\lambda_2 - \lambda_1)\left((h_2^2 - h_1^2)\cos^2\alpha - h_1 h_2 \sin(2\alpha)\right)$$

depends on two scale parameters λ_1, λ_2 and a rotation parameter $\alpha \in [0, 2\pi)$.

Let $\widehat{\gamma}(h)$ be an empirical variogram estimated from the observed data $\{y(u_i)\}$ for the field Y and $\gamma_\beta(h)$ the theoretical parametric variogram model considered in (7) with parameter vector $\beta = (a_1, b_1, c_1, b_2, \lambda_1/c_2^2, \lambda_2/c_2^2, \alpha)$. In practice, only a finite number m of values $\widehat{\gamma}(h_1), \ldots, \widehat{\gamma}(h_m)$ can be computed. In the case of traffic data, the classical least squares method is employed to fit γ_β to $\widehat{\gamma}$. Since traffic data is substantially anisotropic, the variogram model (7) has to be fitted to the data on the whole grid as well as in two directions with polar angles α and $\alpha + \pi/2$.

3.3 Drift Estimation

The mean field $\{m(u)\}$ can be estimated from the data by various methods ranging from radial extrapolation to smoothing techniques such as moving average and edge preserving smoothing. In what follows, the moving average is used because of its ease and computational efficiency for large data sets. By moving average, the value $m(u)$ is estimated as

$$\widehat{m}(u) = \frac{1}{N_u} \sum_{u_i \in W(u)} X(u_i) \tag{9}$$

where $W(u)$ is the "moving" neighborhood of location u and N_u denotes the number of measurement points $u_i \in W(u)$. For fast computation, we put $W(u)$ to be a square with side length τ centered in u.

3.4 Residuals formed with estimated drift

In the previous sections, we supposed that the drift $m(u)$ is explicitly known. However, if it has to be estimated from the data, the theoretical background for the application of the kriging method breaks down (cf. [3], pp. 122–125, [4] p. 72, [6], p. 214). Nevertheless, practitioners continue to use the ordinary kriging of residuals with estimated drift based on the data $y^*(u_i) = x(u_i) - \widehat{m}(u_i)$, $i = 1, \ldots, n$ legitimized by its ease and satisfactory results.

4 Extrapolation of the Velocity Field

In what follows, the extrapolation method of Sect. 3 is applied to real traffic data of the directional sector $S_2 = \{\alpha : \pi/2 \leq \alpha < \pi\}$. This partial data set contains 19699 entries of taxis moving northwest collected over 90 days.

In Fig. 3, the northwest movement direction of the taxis can be clearly recognized in the mean velocity field $\{\widehat{m}(u)\}$. Grey tones reflect speed variation. It clearly shows that the estimator $\widehat{m}$ preserves the spatial velocity structure. To estimate the variogram γ^* of Y^*, the empirical variogram $\widehat{\gamma}_i^*$ is computed for each day $i = 1, \ldots, 90$ and then averaged over all days, i.e. $\widehat{\gamma}^*(h) = \left(\sum_{i=1}^{90} \widehat{\gamma}_i^*(h)\right)/90$. The empirical variogram $\widehat{\gamma}^*(h)$ with "maximum range" in northwest direction and "minimum range" in orthogonal direction is zonally anisotropic; see Fig. 4. The main directions of anisotropy are closely connected to the road directions in downtown Berlin. See Sect. 5 and especially Fig. 10(a) for details.

The zonally anisotropic variogram model (7) with two fixed parameters $\alpha = 170°$, $\lambda_1/c_2^2 = 1000$ has been fitted to the empirical one. The classical least squares fitting method applied to one-dimensional vertical slices of the empirical variogram in orthogonal directions $\alpha = 80°$ and $\alpha = 170°$ yields the remaining parameter values $a_1 = 31.772$, $b_1 = 116.211$, $c_1 = 245388.671$,

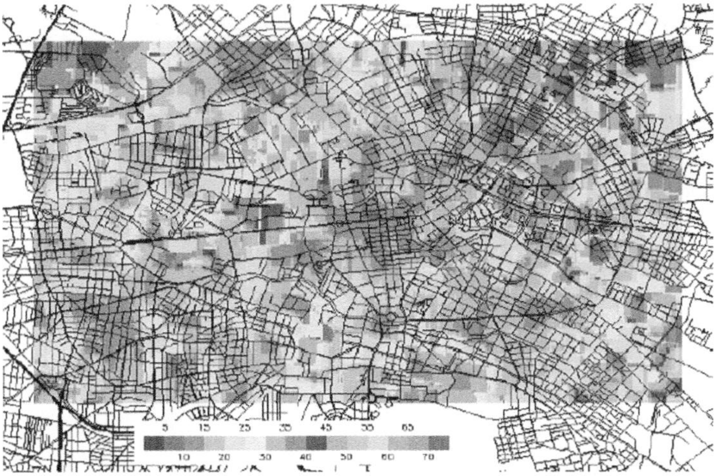

Fig. 3. Mean field $\widehat{m}(u)$ of data set 2

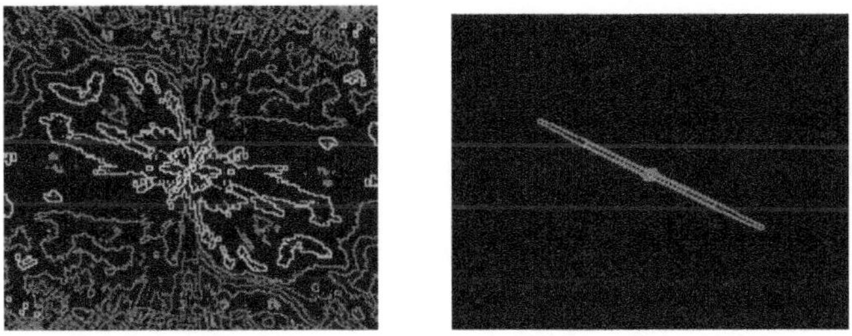

Fig. 4. Empirical variogram $\widehat{\gamma}^*(h)$ and fitted variogram model $\gamma^*(h)$ (*level curves*)

$b_2 = 22.634$, $\lambda_2/c_2^2 = 683964.794$. Thus, the range values in directions 170° and 80° are $r_1 = 270$ m and $r_2 = 162$ m, respectively. It means that the velocities of two vehicles separated by distances $3r_1 = 810$ m in horizontal direction and $3r_2 = 486$ m in vertical direction are almost independent. These range values are conform with the results stated in Sect. 5 for the typical distance between two subsequent positions of the same test vehicle.

For extrapolation, the sample of velocities $x(u_1), \ldots, x(u_n)$ $(n = 223)$ observed on Monday, 18.02.2002 is used. Compared to the whole data set 2 representing the "past", it is interpreted as "actual" data. The random field Y^* of deviations from mean velocities is extrapolated by kriging with moving neighborhood (4) using the indicator function $\mathbf{1}\{u_i \in A(u)\} = \mathbf{1}\{\varphi(u_i - u) \in S_2\}$ where $\varphi(u_i - u)$ is the polar angle of the vector $u_i - u$. This assumption is rather intuitive since only those measurements at locations u_i lying "ahead" of the current position u can influence its velocity value.

The extrapolated residuals $\widehat{Y}^*(u)$ and the resulting velocity map $\{\widehat{X}(u)\}$ are shown in Figs. 5 and 6, respectively. Due to the particular asymmetric

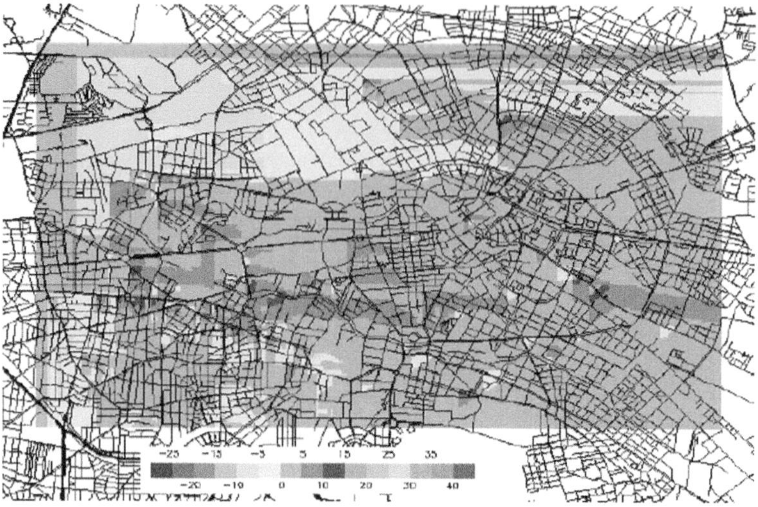

Fig. 5. Residual field $\widehat{Y}^*(u)$

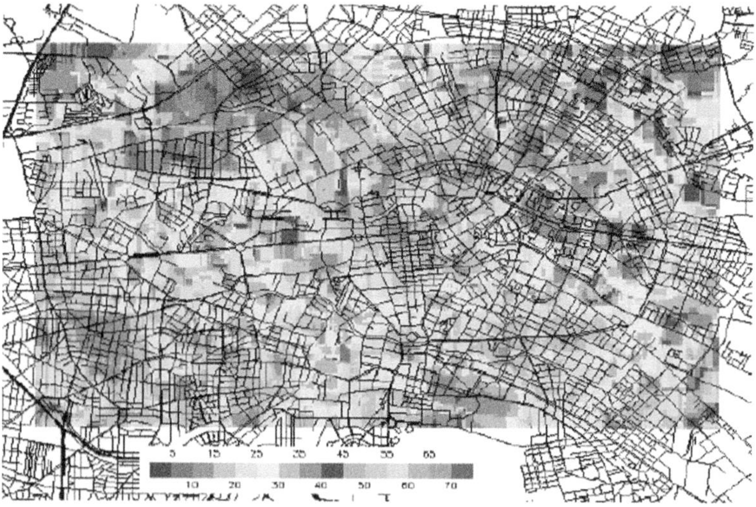

Fig. 6. Velocity field $\widehat{X}(u)$

form of the indicators, the extrapolated field of residuals is strongly discontinuous. Discontinuities of the realizations of $\{\widehat{X}(u)\}$ caused by the kriging with moving neighborhood are essential for precise localization of traffic-jam areas. In Fig. 5, most of the deviation values are zero. The routes of taxis driving in the streets are marked by peaks of the field $\widehat{Y}^*$ with subsequent tails of non-zero residual velocity values lying behind. Thus, one can distinguish separate routes of different test vehicles. See also the extracted taxi routes in Fig. 8, which are similar to those shown in Fig. 5.

Fig. 7. Traffic jams: $\widehat{X}(u) \leq 15$ kph

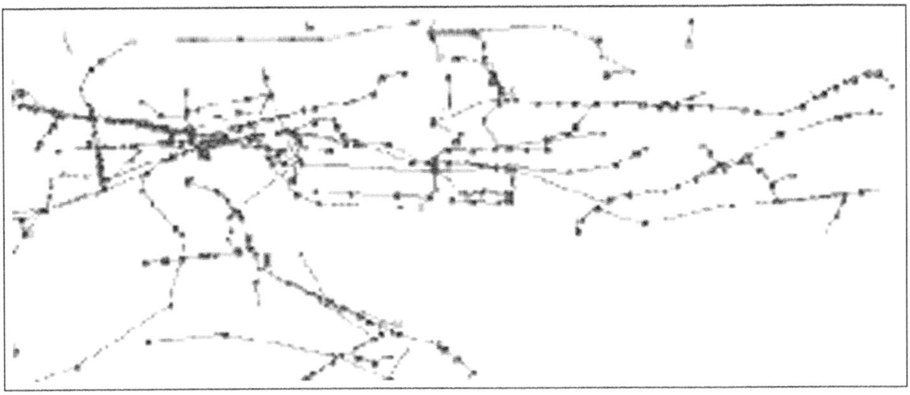

Fig. 8. Taxi routes on 18.02.2002, from 5.15 to 5.30 pm

In Fig. 7, areas with velocities $\widehat{X}(u) \leq 15$ kph are marked grey. Some of these regions might be caused by traffic jams, others are regions with low average velocities. Indeed, the most likely velocity value in downtown Berlin is about 20 kph as it can be seen in Fig. 12.

5 Statistical Analysis of Traffic Data

In addition to the spatial statistical inference performed above, we now discuss the histograms of velocity residuals and further traffic characteristics which bring an extra insight into the structure of traffic data. In particular, they help us to explain some features of anisotropy and spatial correlation which we already mentioned in Sect. 4. For a more detailed treatment of the subject, see [2].

5.1 Distributional Properties of Polygonal Taxi Routes

If we think about the way the traffic data are collected we understand that the locations where the velocities are measured can not be deterministic. Moreover, they are stochastically dependent. In fact, each test vehicle follows a route that consists of a random number of segments. Each segment connects two locations where consecutive GPS signals were sent; see Fig. 8. The histogramm of the number of segments in the taxi routes is shown in Fig. 9. It turns out that this histogram can be well approximated by a geometric distribution with parameter $p = 0.9365064$ being the probability of enlarging a route by a new segment. Furthermore, the geometry of the taxi routes explains the form of the variogram anisotropy mentioned in Sect. 4.

In particular, the distribution of the angles between the movement direction of a vehicle and the eastward direction in Fig. 10(a) reflects the distribution of typical street directions with heavy traffic in downtown Berlin. The majority of main roads goes east or west which corresponds to the angles of $0°$, $180°$, and $360°$, respectively. This is certainly the reason for the character of zonal anisotropy of the variograms in Fig. 4. Figure 10(b) shows that the distribution of segment lengths is demonstrably non-normal. Furthermore, with probability of ca. 0.9, the distances between two subsequent GPS signals in the taxi routes do not exceed 1000 m. It is clear that the velocities at two positions within this distance are correlated. The opposite statement is also true. As it has been already mentioned in Sect. 4, the velocities of two cars

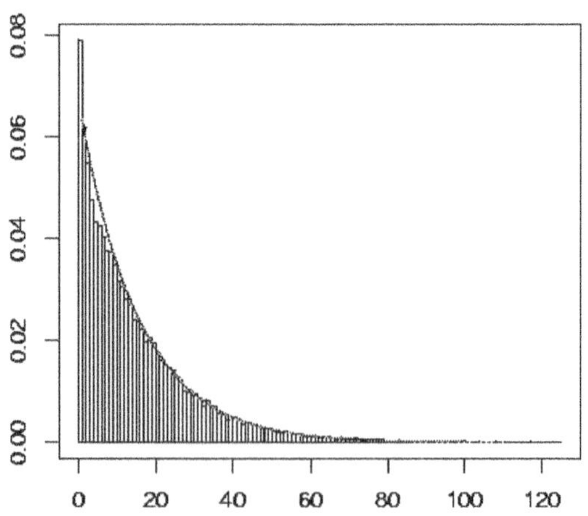

Fig. 9. Histogram of the number of segments in the taxi routes

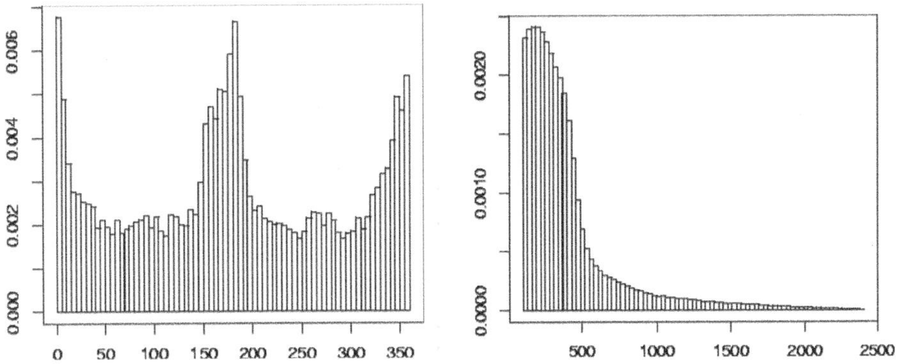

(a) Directions of route segments
(b) Lengths of route segments

Fig. 10. Histograms of segment directions (in degrees) and lengths (in m)

at a distance of more than $3\sqrt{r_1^2 + r_2^2} \approx 945$ m from each other are almost independent.

5.2 Distribution of Velocity Residuals

The histogram in Fig. 11 shows that the distribution of velocity residu als can be well fitted by some normal distribution. Nevertheless, a more detailed statistical inference shows that the distribution of velocity residuals depends on the value of mean velocity. One reason for this is that the sum of the residual and the mean has to be non-negative. Figures 13, 14 and 15 show the histograms of the velocity residuals measured at locations with mean velocities (in kph) belonging to three disjoint classes: $[15, 20)$, $[25, 30)$ and $[40, 45)$, respectively.

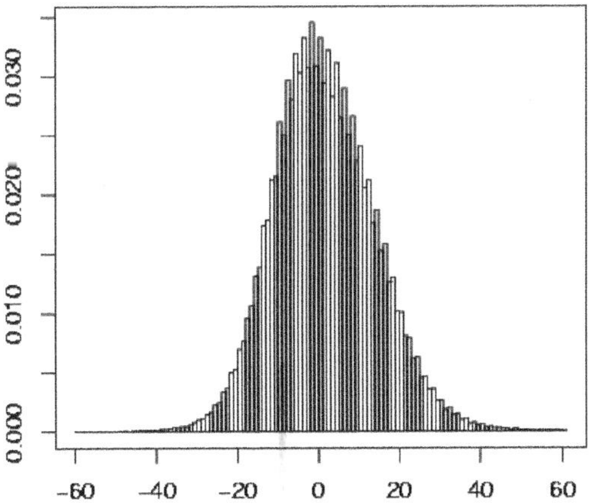

Fig. 11. Histogram of the velocity residuals (in kph)

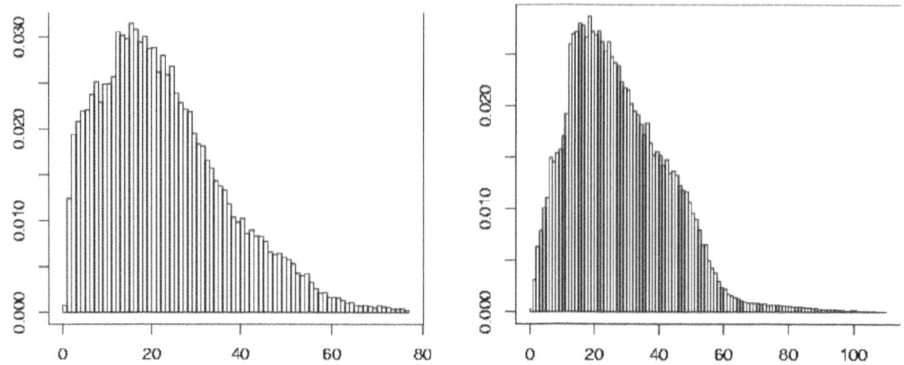

(a) for the first route segments (b) for the remaining route segments

Fig. 12. Histogram of velocities (in kph)

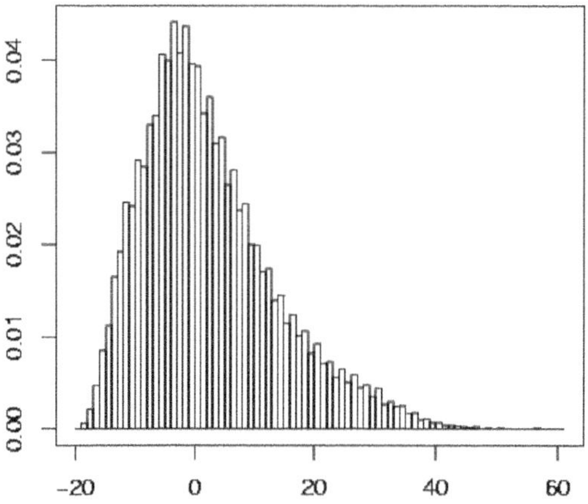

Fig. 13. Histogram of residuals given that $\widehat{m} \in [15, 20)$

The right skewness of this histogram means that large positive deviations from small mean values are more likely than negative ones.

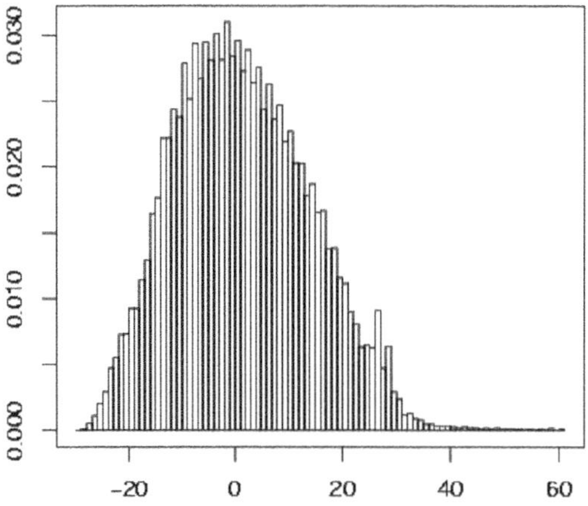

Fig. 14. Histogram of residuals given that $\widehat{m} \in [25, 30)$

This histogram is almost symmetric. Thus, both positive and negative residuals of equal size occur nearly with the same probability.

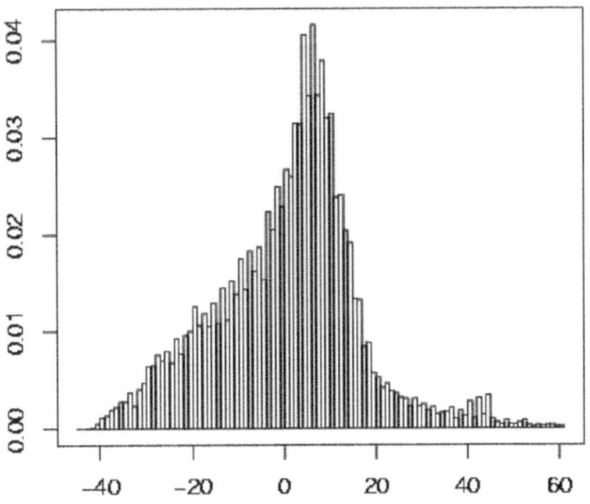

Fig. 15. Histogram of residuals given that $\widehat{m} \in [40, 45)$

It is clear from this histogram that at positions with large mean values large negative residuals are more likely than positive ones.

If we add the mean velocity values to their residuals we see that most velocities in downtown Berlin do not exceed 60 kph. The histogram of the velocities themselves is given in Fig. 12 which shows that the most likely velocity value in downtown Berlin (i.e., the modus of the empirical velocity distribution) is about 20 kph. This explains the dominance of low velocity values in the mean field $\widehat{m}$ and the threshold maps in Figs. 3 and 7.

6 Outlook

The spatial extrapolation and statistical space-time analysis of traffic data considered in the present paper is an important step towards stochastic modelling, simulation and prediction of future road-traffic states. Our results can be used to construct a Markov-type simulator by means of which future routes of test vehicles can be generated, where the choice of the starting configuration depends on the actually measured traffic situation. In particular, when sampling the velocity residuals from histograms as given in Figs. 13, 14 and 15, the mean velocity field $\{\widehat{m}(u)\}$ will be actualized by the recently, say at the given day, observed velocities. For example, suppose that significantly larger velocities than usually have been observed at the considered day in a certain neighborhood of location u. In this case, the velocity residual at location u will be sampled from a histogram which corresponds to a larger class of mean velocities than the "historical" value $\widehat{m}(u)$. Further details concerning our simulation algorithms can be found in [2].

Then, using the extrapolation technique described in Sects. 2 and 3, velocity maps based on both the measured and simulated traffic data can be computed. To evaluate the quality of these maps, they are compared with corresponding velocity maps computed exclusively from measured traffic data. The comparison is based on morphological distance measures for digital image data. These issues will be discussed in a forthcoming paper.

Acknowledgement

This research was supported by the German Aerospace Center (DLR) through research grant 931/69175067. The authors are grateful to Reinhart Kühne, Peter Wagner and their co-workers from the DLR Institute of Transport Research for suggesting the problem as well as for stimulating and fruitful discussions on the subject.

References

1. BRAXMEIER H, SCHMIDT V, SPODAREV E (2004) Spatial extrapolation of anisotropic road traffic data. *Image Analysis and Stereology* 23, 185–198.
2. BRAXMEIER H, SPODAREV E, SCHMIDT V (2005) Statistische Raum-Zeit-Analyse und Simulation von Verkehrsströmen in Ballungsgebieten. Verkehrsforschung - Online 2, http://www.verkehrsforschung-online.de
3. CHILÈS JP, DELFINER P (1999) *Geostatistics: modelling spatial uncertainty.* Wiley, New York
4. CRESSIE NAC (1993) *Statistics for spatial data.* Wiley, New York
5. GEOSTOCH (2008) Java library. University of Ulm, Department of Applied Information Processing and Department of Stochastics, http://www.geostoch.de
6. WACKERNAGEL H (1998) *Multivariate geostatistics.* 2nd ed., Springer, Berlin

On Evaluation of Precipitation Fields with Rain Station Data

Bodo Ahrens

Institut für Meteorologie und Geophysik, Universität Wien, Wien, Austria
Bodo.Ahrens@univie.ac.at

1 Introduction

Nowadays, limited area numerical weather prediction models provide meteorological forecasts with horizontal grid spacing of only a few kilometers and grid spacing will decrease further in the coming years caused by progress in high-performance computing [9, 26]. Precipitation forecasts are of primary interest for both researchers and the public. For example, in flood forecasting systems precipitation is the crucial input parameter, especially in mountainous watersheds. Like the grid spacing of weather prediction models the grid spacing of regional climate models is decreasing.

Precipitation forecasts have to be evaluated and errors have to be quantified. The most important evaluation method is comparison of meteorological simulation results with meteorological observations. But, before errors can be quantified two decisions have to be made. First, a set of useful statistics has to be chosen. This shall not be the issue of this paper. The interested reader is referred to, for example, Murphy and Winkler [23], Wilks [31], Wilson [32]. We apply for illustration a small set of simple continuous statistics.

Our focus is on the second problem: What is the observational reference? Rain station data is commonly preferred to remote sensing data, in particular radar data, because of the relatively large measurement uncertainties [e.g., 1, 14, 34]. Is it reasonable to compare precipitation forecasts valid for grid boxes with several kilometers in diameter with sparsely distributed rain station data valid for small areas of $\sim 1000\text{cm}^2$? This is often done in an operational framework since it can be implemented by simple means. This area-to-point evaluation is criticized and it is proposed to perform some upscaling or regionalization of the station data up to forecast grid resolution [29, 12]. Regionalization can be done by some fitting approach yielding a precipitation analysis. For example, a recent analysis of precipitation for the European Alpes by Frei and Häller [17] has a time resolution of 24 h and a spatial grid of about 25 km with regionally even lower effective resolution depending on the available surface station network. This type of analysis is useful for model validation at the 100 km-scale [see, e.g., 5, 16, 18], but not at 10 km-scale or even less.

Analysis is a smoothing regionalization. This deteriorates application in higher-moment statistics if the network is not dense enough. The statement "dense enough" critically depends on the applied pixel support (is a pixel value representative for boxes with diameter of ~100, 10, or 1 km?) and the analysis scheme. Another regionalization approach is stochastic simulation of precipitation fields with conditioning on the available station data. The idea of this is that the data is respected and the spatial variability is represented more realistically than in the analysis. Then the forecast can be compared with an ensemble of simulated fields. The ensemble mean field is an analysis but the mean higher-moment statistics have not the same value than if the forecast is just compared with the analysis alone.

This paper applies regionalization and performs area-to-point or area-to-area comparison in evaluation of daily precipitation forecasts. The forecasts to be evaluated by example are the forecasts of the NWP model ALADIN that is operational at the Austrian national weather service with 10 km grid spacing. ALADIN, the forecast days, and the available station data are introduced in the next section. Section 3 discusses the applied evaluation approaches and subsequent sections discuss the respective results. Finally, some concluding remarks will be given.

2 Precipitation Events and Data

For illustrational purposes we investigate the August 2002 flood in Austria. It was caused by two devastating, large scale rain events. The first rain event (6–8 August) was dominated by an upper-air low over central Europe. This flow pattern lead to torrential rainfall, especially on the windward-side of the northern Alps and along the Austrian–Czech border. The second rain event (11–13 August) was caused by strong cyclogenic activity over the Mediterranean sea resulting in a south-easterly flow of warm and moist air into Central Europe. Here, the daily precipitation fields over Austria for August the 6, 7, 11, and 12th are of interest. An investigated day starts at 06 UTC and finishes at 06 UTC the next day.

Observational rain data is available from two sets of rain station data. The first set is provided by the *Hydrologische Zentralbüro* with about 800 stations. This set is named HZB in the following. The second data set is provided by the Austrian national weather agency ZAMG with about 116 stations measuring during the 4 days of evaluation. This second data set, named TAWES, is independent from the HZB data set and generated by automatic weather stations and available in near real time. Within this paper the daily time scale is applied. Thus the TAWES data is accumulated to daily values.

This paper compares observational precipitation data with forecast precipitation fields from a numerical weather prediction (NWP) model. This comparison is a crucial step of model evaluation. Here, we apply forecast fields simulated with the NWP model ALADIN (Aire Limitée Adaptation Dynamique dévelopement InterNational, see, e.g., Bubnova et al. [11], Ahrens et al. [4] and http://www.cnrm.meteo.fr/aladin/) in the setup operational in the year 2002 by the Austrian national weather service with about 10 km horizontal grid spacing. Thus, the precipitation forecast values are valid for $10 \times 10\,\mathrm{km}^2$ blocks. Evaluated are sequences of 30-h forecasts initialized at 00 UTC discarding the leading 6 hours to account for model spin-up. Previous investigations have already shown that ALADIN quantitative precipitation forecasts are useful [e.g., 2, 21, 22].

3 Evaluation Methods

In the following we will discuss the evaluation procedures applying a minimal set of useful statistics. Most important is the mean distance bias $= 1/N_x \sum_{x=1}^{N_x}(m_x - d_x)$ with the model forecast field m_x, the observational field d_x, and with the space index $x = 1, \ldots, N_x$. Additional statistics considered are the coefficient of determination R^2 (the square of the linear product-moment correlation, possible values are between 0 and 1 with optimal value 1), and the ratio of spatial variances SPREX $= \sigma_m^2/\sigma_d^2$ (optimal value 1).

The applied evaluation methods are comparisons of model fields with (a) station data (i.e., effectively with point data) and (b) with regionalized and box averaged precipitation fields (i.e., with area data). Comparison with point data is often done and, for example, standard in most European Meteorological Services [see 10, 32]. It is simple in implementation. Two variants are common practice: direct comparison of the station data with the closest model grid box values and thus performing an area-to-point comparison, or interpolation of the model fields to the station locations and thus performing a point-to-point comparison. In fact this interpolation smoothes the forecast field that is eligible since single box values should not be interpreted [2, 20]. But, on the other hand a simple interpolation like the often applied bi-linear interpolation assumes that the precipitation field is continuous and introduces no additional information. Consequently, interpretation of the interpolated values as point data is delusive. The effective resolution of ALADIN is not the issue here, and the raw forecasts of ALADIN with about 10 km horizontal resolution are evaluated by example.

Comparison of model grid box output with regionalized rain fields with appropriate pixel support is an area-to-area comparison and respects the scales. The second potential advantage of regionalization is that station representativity problems (clustering of stations around larger cities or along valleys) can principally be compensated. Here, we call regionalization by some optimization involving data-fitting techniques (like regression, polynomial fitting, spline functions, kriging, etc.) analysis and the estimated field is an analysis field. A problem of analysis is that it is difficult to estimate analysis errors (e.g., Kriging variances underestimate the analysis error in case of precipitation since the Kriging assumptions are not fulfilled). A second problem is that analysis fields are expected to underestimate the true field variance (e.g., the smoothing relationship of Kriging states that the analysis variance at any location is the data variance minus the kriging variance). These problems have to be taken into account in an evaluation investigation.

Here, the details of the analysis scheme are of minor importance and two schemes are applied. First, ordinary block Kriging with spherical variogram model and, second, inverse squared-distance weighting interpolation with blocking is applied. Chosen blocks are 10 km in diameter and thus pixel support of the analysis is $10 \times 10\,\mathrm{km}^2$ like of the NWP model forecasts. In case of HZB data analysis a neighborhood of 64 stations and in case of TAWES data of 8 stations is considered. Kriging variants are often proposed and applied in precipitation analysis [7, 8, 15, 19]. Tests have shown that ordinary Kriging with spherical variogram, whose parameters are estimated from the actual data, performs slightly better than with other tested variograms or with universal Kriging in the events investigated. Figure 1 shows the block kriging results with pixel support of 10 km for the four days of comparison. Figure 2 shows the model forecasts. Obviously, the model performance is worst at day one and best at day four. In the following we discuss the quantification of this subjective conclusion.

Another regionalization approach is stochastic simulation. Here, the usefulness of precipitation field simulation conditioned on available data shall be shown by application of Gaussian conditional simulation [e.g., 13, Chap. 7] with 10 km blocks. The conditioning respects the station values and the estimated variograms.

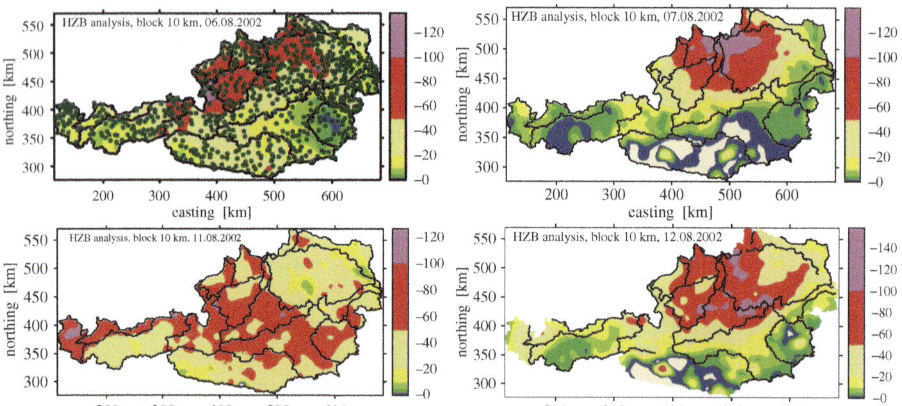

Fig. 1. Daily analyses by ordinary block Kriging of HZB rain station data in Austria for the four days investigated in this paper. The *dots* in the *upper left panel* indicate the station locations. Units are mm/d

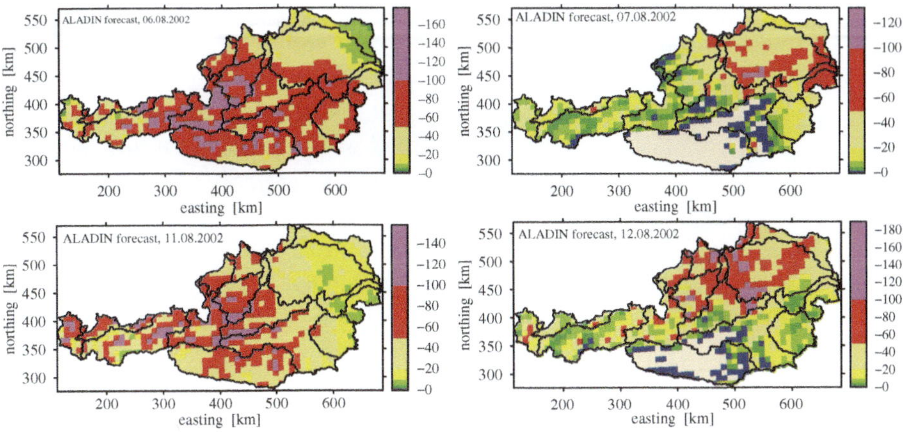

Fig. 2. Operational NWP model forecasts for the days shown in Fig. 1. Units are mm/d

Precipitation is a non-Gaussian, non-negative process and, therefore, the chosen simulation method is far from optimal. There are unconditional simulation methods described in the literature [e.g., 3, 30], but there is no appropriate conditional approach known to the author. Additionally, normal score transformation of the precipitation data has not been proven superior to direct simulation. Here, the advantages of regionalization by simulation shall be discussed and the mentioned deficiencies are not crucial for the presented conclusions. The made assumptions are the same as in Kriging. The statistics calculated with the simulated fields are determined by the raw data, i.e. if simulated with negative precipitation data. In Fig. 4 (negative values set to

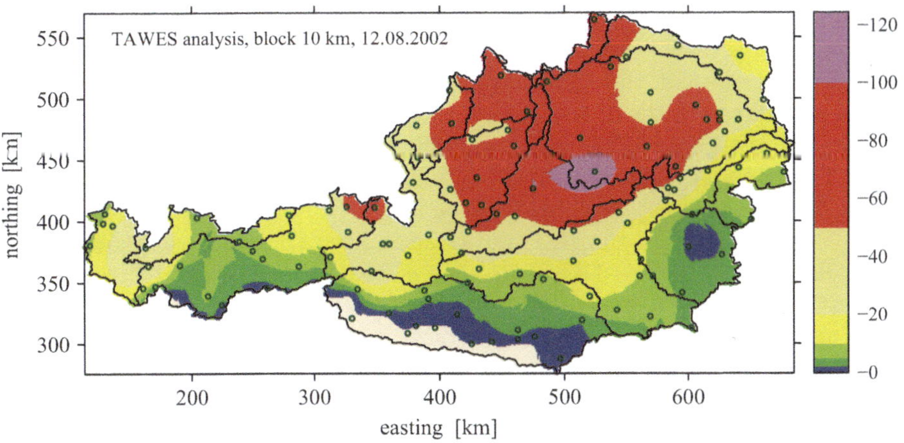

Fig. 3. Daily analysis by ordinary block Kriging of TAWES rain station data. Units are mm/d

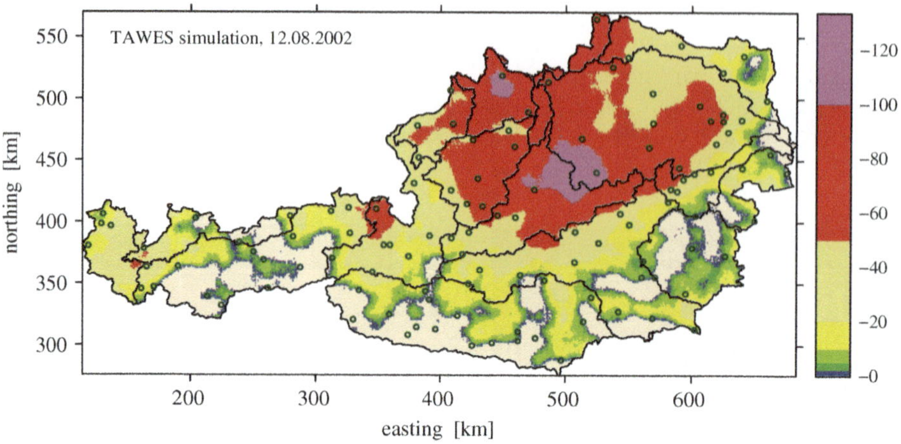

Fig. 4. Realization of a conditional simulation to the daily accumulated TAWES data applied in Fig. 3. Units are mm/d

zero) an example of an simulated precipitation field conditioned on TAWES data is compared with the Kriging analysis of these data. The simulated field is more heterogeneous with slightly larger maximum values. It is assumed that a more realistic simulation approach would realize even more variable fields.

Kriging, inverse distance weighting interpolation and Gaussian simulation are performed with the geostastistical software package *gstat* [24, and www.gstat.org in the World Wide Web].

4 Evaluation Against Station Data

In a first evaluation step the comparison of model grid box data with rain station data is discussed. The nearest model grid box is used to compare with the point observations ignoring the corresponding error in location. Figure 5 and Table 1 present the comparison with the HZB and TAWES data set. There is a large scatter in the results depending on the applied reference: the TAWES or HZB data set. For example, the relative bias is +4% in comparison with TAWES and −4% in comparison with HZB data at August 7th. At the same day the forecast explains 39% of the TAWES data variability ($R^2 = 0.39$) but only 20% of the HZB data variability. The model underestimates the field variance at the 11th if compared with HZB data by 10% or rather overestimates by 10% in comparison with TAWES data.

The impact of the station sample size is illustrated by the box plots in Fig. 5. Twenty random sub-samples of 116 stations (the sample size of the TAWES set) are drawn from the HZB set. These sub-samples are applied in the evaluation process and the box plots show the quartiles of the twenty evaluation results for each day and statistics. Twenty is a small number of random sub-samples, but enough to illustrate the effects. The range of these

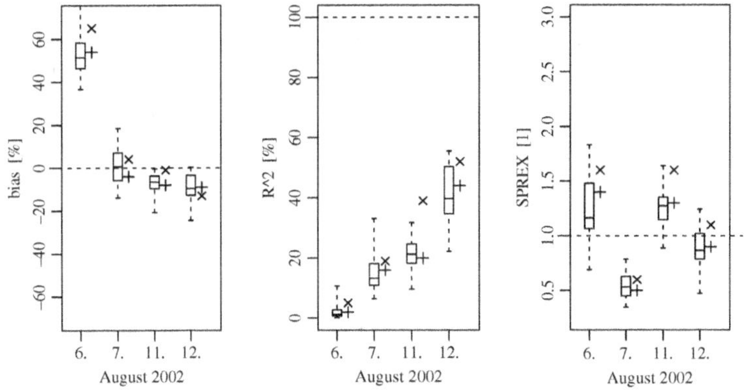

Fig. 5. Comparison of NWP model forecasts against station data with the symbols "+" indicating the evaluation statistics of forecast vs HZB station data for the four days and with "×" indicating the comparison results against the TAWES data. The box plots show the quartiles (the whiskers indicate the range) of results of comparison against 20 random subsets with sample size 116 from the HZB data

Table 1. Comparison of forecasts by the NWP model against HZB or TAWES station data by statistics. The given values are valid for August 6/7/11/12, respectively

	bias [%]	R^2 [%]	SPREX [1]
NWP∼HZB	54/ − 4/ − 8/ − 9	2/16/20/44	1.4/0.5/1.3/0.9
NWP∼TAWES	65/4/ − 1/ − 13	5/19/39/52	1.6/0.6/1.6/1.1

results is substantial. For example, the relative bias range is about 20% for the days with small bias and even 43% for August 6th. Interestingly, the TAWES results are not within the interquartile range of the sub-sampling results most of the time and the difference is systematic (besides the bias at August 12th). The extremness can be explained by a more homogeneous distribution of the TAWES stations (cf. Fig. 3) in comparison with the sub-sampled HZB stations and more important by the different measurement system. The problem of rain measurements can not be discussed further and the interested reader is referred to [25, 33].

Instead of the next model box value often a bi-linearly interpolated value is compared with station observations as discussed above. The bias results are different to the results with next neighbor comparison, but within a scatter range suggested by the box plots. The results of the pattern comparison improves slightly but systematically (up to 5%). This is not surprising since the model data is implicitly smoothed by the interpolation and thus the classic "double-penalty problem" [small location discrepancies of sharp peaks are penalized twice, cf. 6] is reduced. Smoothing reduces the model variance and thus the values of SPREX decrease (by 0.05 for the 7th to 0.2 for the 12th in case of TAWES data).

5 Evaluation Against Analyses

With analyzed data it is possible to compare NWP forecast fields with analysis fields at the same scale (pixel support, grid spacing, and coverage). Here, as mentioned above, we apply ordinary block Kriging (OK) and inverse distance weighing interpolation (IDW) with a block-size comparable to the box area of the forecast model of $10 \times 10\,\mathrm{km}^2$.

Table 2 shows values of the statistics of comparison of forecast fields (NWP) with analyses and intercomparisons of analyses. The scatter in the statistics' values due to analyzing the HZB or TAWES set in NWP model evaluations decreases slightly in comparison with evaluation against station data. This is supported by Fig. 6. The box-plots show the quartiles of values from twenty comparisons of analyses by OK with the forecasts and the relative range in the statistics is smaller (at the 11th the absolute range increased but the median approximately doubled).

Table 2. Comparison of NWP forecast fields with different analyses and of analyses with analyses. Analyses considered are done by Kriging (OK) or inverse distance weighting interpolation (IDW) based on different sets of rain station data (TAWES or HZB). The last row shows the mean statistics of twenty comparisons of analyses based on HZB sub-sets versus the total HZB data set

	bias [%]	R^2 [%]	SPREX [1]
$\mathrm{NWP} \sim \mathrm{OK_{HZB}}$	$73/-14/-0/-11$	2/23/21/47	1.7/0.5/2.2/1.0
$\mathrm{NWP} \sim \mathrm{OK_{TAWES}}$	$70/-14/-1/-8$	1/25/30/51	1.8/0.5/2.7/1.1
$\mathrm{NWP} \sim \mathrm{IDW_{HZB}}$	$73/-16/1/-12$	2/23/23/50	2.1/0.5/3.1/1.2
$\mathrm{NWP} \sim \mathrm{IDW_{TAWES}}$	$71/-11/-2/-8$	1/24/32/54	2.0/0.5/3.2/1.3
$\mathrm{OK_{TAWES}} \sim \mathrm{OK_{HZB}}$	$2/0/0/-4$	84/93/62/87	0.9/1.0/0.8/0.9
$\mathrm{IDW_{TAWES}} \sim \mathrm{OK_{HZB}}$	$2/-4/2/-3$	84/92/60/83	0.8/0.9/0.7/0.8
$\langle \mathrm{OK_{SS}} \sim \mathrm{OK_{HZB}} \rangle$	$-2/2/-1/1$	79/93/55/81	0.8/1.0/0.8/0.9

There are differences if NWP is compared against IDW or OK analyses. The values for R^2 and SPREX are larger with IDW since the IDW analyses are smoother than the OK analyses as is quantified by SPREX in row 5 and 6 of Table 2. This is reasonable since the influence of distant data is described in OK by the applied variogram that increases close to the origin faster than quadratic which is assumed in IDW. The effective pixel support is larger for IDW analysis than for Kriging analysis. Thus, even if formally the pixel support is appropriate (both analyses are estimated for $10 \times 10\,\mathrm{km}^2$ blocks) the second moment statistics are sensitive to the effective smoothness (cf. analysis vs analysis SPREX values in Table 2 or the Fig. 3 in comparison with the lower-right panel of Fig. 1). Keeping this in mind the area-to-point comparison of the last section is not fair to the forecasts: R^2 and SPREX are too small in the mean.

Since the assumptions of Kriging are not fulfilled very well by precipitation data it is not sure a-priori that Kriging is superior to IDW. Nevertheless, the impact of the analysis scheme is smaller than the impact of data sample size. This is evident if, for example, the coefficient of determination is more closely inspected for the third analysis day, August 11th. Both analyses based on TAWES data explain only about 60% of the variance of the HZB analysis. This day is less dominated by large-scale precipitation patterns and shows the smallest field variance, but has the largest small-scale variability. Small shifts in the analyzed small-scale pattern lead to more distinct double-penalty effects than for the other days. And these shifts are less influenced by the analysis method as the similar R^2 values indicate than by the smaller data sample size. This conclusion is supported by the last row in Table 2 where the mean error of analyses based on random HZB subsets with sample size 116, as of the TAWES set, is shown. Nevertheless, a value of 60% is enough if compared to the value of 21% explained by the NWP. What is missing is a possibility to judge these 21% in comparison to the 23% for the 7th where the precipitation field is far more easy to analyze as the more than 90% explained variance indicates and thus should also be easier to forecast in the sense of the applied statistics.

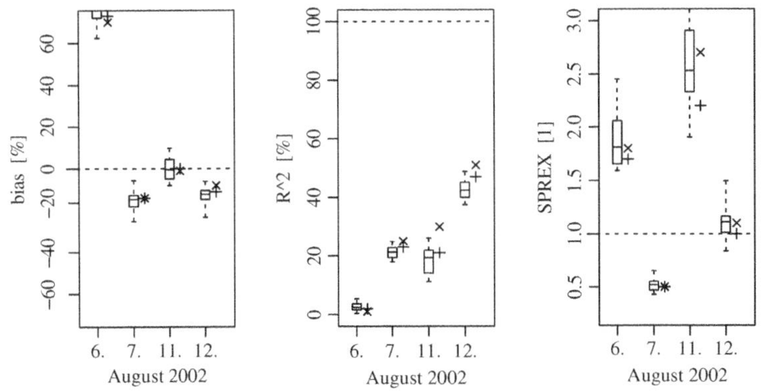

Fig. 6. Comparison of NWP model forecasts against analyses. The symbols "+" indicate the evaluation statistics of forecast vs HZB based analysis for the four days and "×" indicates the results vs TAWES based analysis. The box plots show the quartiles (the whiskers indicate the range) of results of comparison against 20 analyses based on random HZB subsets with sample size 116

Table 3. Statistics' mean of comparisons of NWP forecasts with twenty simulations conditioned on TAWES data and of the simulations with the HZB or TAWES analyses

	bias [%]	R^2 [%]	SPREX [1]
$\langle NWP \sim SI_{TAWES} \rangle$	69/ − 13/ − 2/ − 11	1/25/24/48	1.6/0.5/2.2/1.0
$\langle SI_{TAWES} \sim OK_{HZB} \rangle$	2/1/1/ − 2	72/85/48/70	1.0/1.1/1.0/1.1
$\langle SI_{TAWES} \sim OK_{TAWES} \rangle$	−0/0/1/2	84/91/77/90	1.2/1.1/1.2/1.2

6 Evaluation Against Simulations

Analysis underestimates spatial variability. The idea of stochastic simulation is that the stochastic realizations show a good representation of the natural field variability. Here, as mentioned above, Gaussian simulation with 10 × 10 km blocks is applied. This is a oversimplified approach, but the principal advantages can be discussed.

Figure 4 shows a realization of a simulation conditioned on TAWES data that is obviously more variable than the corresponding analysis. Consequently, R^2 (due to the double-penalty effect) and SPREX of NWP in comparison with simulated fields (Table 3) are smaller in the mean than in comparison with the TAWES analysis (Table 2) and closer in the mean to the comparison with the HZB analysis. This can also be seen in Fig. 7 through the systematic shifts of the box plots from the TAWES to the HZB analysis values. In case of SPREX the interpretation is that the natural variability is indeed more realistically represented in the simulations than in the TAWES analysis. In case of R^2 the double penalty effect is larger in the realizations than in the TAWES analysis. Since the simulation R^2s are rather close to the HZB R^2s we conclude that the pattern at scales finer than the effective TAWES analysis resolution are not forecast.

Also exciting are the comparisons of the simulated realizations against the analyses in Table 3. In the mean the TAWES simulations explain only about 50% of the HZB analysis variance for August 11th, but more than 70% for the other days. The 11th is also the day with smallest mean R^2 if the simulated fields are compared with the TAWES analysis. Therefore, the available TAWES data are a relatively small constraint for the simulations for this day, and the regionalizations, the simulations or the analysis, and thus the evaluation results for this day are most uncertain. Additionally, small-scale variability is important that is generally not very well forecasted by NWP

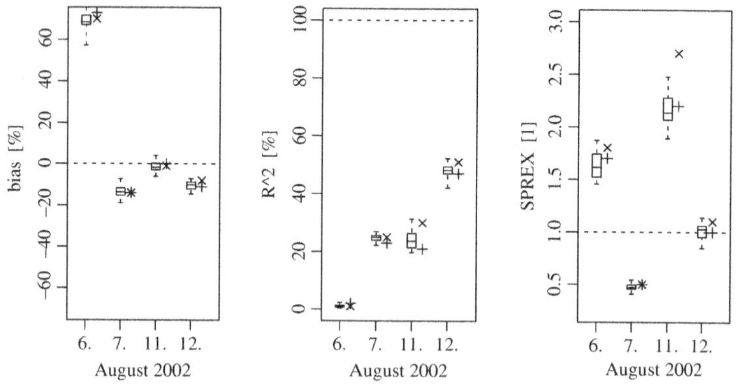

Fig. 7. Same as Fig. 6, but the box plots summarize the comparison of NWP forecasts against precipitation simulations conditioned on TAWES data

models due to predictability constraints. Thus, the forecast of August 11th is more valuable than the forecast of the 7th. The simulations complement the evaluation against regionalized precipitation fields.

7 Conclusions

Precipitation forecasts by numerical weather prediction or climate model precipitation forecasts have to be evaluated. Often this is done on the basis of precipitation rain station data. This paper investigates the effect of different scales and representativity of the compared data. As an evaluation test bed the NWP model ALADIN with 10 km gridspacing is evaluated for four heavy precipitation days in August 2002 by comparison with precipitation data measured with about 800 stations operated by the Austrian hydrological service (HZB data set) and with a independent set of 116 stations operated in near real-time by the Austrian weather service (TAWES data set). Thus, about 100 stations are available in Austria (total area: 84 000 km^2) for a day-to-day evaluation procedure. Somewhat less namely 72 quality controlled stations are available in Austria for a climatic period of time (1948–2002, cf. Schöner et al. [27]).

The experiments show that the uncertainty in bias evaluation is of the order of the bias if the forecasts are compared with a station data set of sample size 116. This scatter can be reduced by best estimate regionalization, named analysis, of the data. A Kriging variant and inverse distance weighting interpolation are applied. The dependence on the analysis scheme is smaller than on data reduction by about a factor of eight.

It is shown that direct comparison of forecast fields with $10 \times 10\,km^2$ pixel support to station data with about $0.1 \times 0.1\,m^2$ is inappropriate if second-moment statistics shall be evaluated. This is exemplified by calculation of a measure for pattern matching, R^2, and a ratio of variances SPREX. Comparison with analyses based on the reference station data set HZB and the smaller data set TAWES proved the positive impact of regionalization by analysis with $10 \times 10\,km^2$ analysis blocks. Nevertheless, there is scatter in the results and, especially, the values for SPREX show that analysis is a smoothing process that may not provide a good representation of the variability, particularly in regions with sparse observations coverage. Here, regionalization by stochastic simulation conditioned on the station data complements the interpretation of the evaluation results. The idea of stochastic simulations is that they represent the natural field variability more realistically. This is confirmed by comparison of TAWES analyses and simulations conditioned on TAWES data with the HZB analyses which are the available reference.

Therefore, the following recipe for forecast evaluation with rain station information is proposed: (1) compare with analyzed fields at the same scale, and (2) apply stochastic simulation conditioned on the data. Step one is applied, for example, in Ahrens et al. [5], Ferretti et al. [16], Frei et al. [18] at grid-scales of about 100 km in space and 24 h in space with analyses based on denser data than assumed here to be available in near real-time. At these relatively coarse scales the analyses represent the first and second moments of the precipitation field with sufficient accuracy. At finer scales with even less data like in the setup assumed here analyses are insufficient, but can be complemented by stochastic simulations. Stochastic simulation is an efficient method for detection and eventually avoidance of difficulties with analyses.

This paper discusses spatial representativity of rain station data in evaluation of NWP or regional climate model forecast. Here, only the horizontal representativity issue is considered. In the mountainous areas the inhomogeneous distribution of stations in the vertical (most stations at valley floors) can lead to systematic errors that are difficult to consider [e.g., 28]. A further problem is systematic errors due to wind and evaporation loss of the rain gauges [e.g., 25]. These are additional difficulties which should be considered, but are often neglected in precipitation evaluation.

Acknowledgements

Precipitation data are provided by the Abteilung Wasserhaushalt im BMLFUW, Hydrologische Zentralbüro, Vienna, and the Austrian weather service ZAMG, Vienna. I thank the maintainers and developers of the statistical software system *R* and especially E. Pebesma and R. Bivand for the development of the software packages *gstat* and *R/GRASS*, respectively. The author is supported by the Austrian Academy of Sciences, Project HÖ 22.

References

1. R. F. Adler, C. Kidd, G. Petty, M. Morissey, and H. M. Goodman (2001) Intercomparison of global precipitation products: The third precipitation intercomparison project (PIP-3). *Bull. Amer. Meteor. Soc.*, 82(7):1377–1396.
2. B. Ahrens (2003) Evaluation of precipitation forecasting with the limited area model ALADIN in an alpine watershed. *Meteorol. Z.*, 12(5):245–255.
3. B. Ahrens (2003) Rainfall downscaling in an alpine watershed applying a multiresolution approach. *J. Geophys. Res.*, 108(D8):8388–8399.
4. B. Ahrens, K. Jasper, and J. Gurtz (2003) On ALADIN precipitation modeling and validation in an Alpine watershed. *Ann. Geophysicae*, 21:627–637.
5. B. Ahrens, U. Karstens, B. Rockel, and R. Stuhlmann (1998) On the validation of the atmospheric model REMO with ISCCP data and precipitation measurements using simple statistics. *Meteorol. Atmos. Phys.*, 68:127–142.
6. R. A. Anthes (1983) Regional models of the atmosphere in middle latitudes. *Monthly Wea. Rev.*, 111:1306–1335.
7. P. M. Atkinson and C. D. Lloyd (1998) Mapping precipitation in Switzerland with ordinary and indicator kriging. *J. Geogr. Inf. Dec. Anal.*, 2(2):65–76.
8. A. Beck and B. Ahrens (2004) Multiresolution evaluation of precipitation forecasts over the European Alps. *Meteorol. Z.*, 13:55–62.
9. R. Benoit, C. Schär, P. Binder, S. Chamberland, H. C. Davies, M. Desgagné, C. Girard, C. Keil, N. Kouwen, D. Lüthi, D. Maric, E. Müller, P. Pellerin, J. Schmidli, F. Schubiger, C. Schwierz, M. Sprenger, A. Walser, S. Willemse, W. Yu, and E. Zala (2002) The real time ultrafinescale forecast support during the special observing period of the MAP. *Bull. Amer. Meteor. Soc.*, 83(1): 85–109.
10. P. Bougeault (2003) The WGNE survey of verification methods for numerical prediction of weather elements and severe weather events. WMO Report, available at http://www.bom.gov.au/bmrc/wefor/staff/eee/verif/verif_web_page.html
11. R. Bubnova, G. Hello, P. Bernard, and J.-F. Geleyn (1995) Integration of the fully elastic equations cast in the hydrostatic pressure terrain-following coordinate in the framework of the ARPEGE/Aladin NWP system. *Monthly Wea. Rev.*, 123:515–535.

12. T. Cherubini, A. Ghelli, and L. Francois (2002) Verification of precipitation forecasts over the alpine region using a high-density observing network. *Weather Forecast.*, 17(2):238–249.
13. J. Chilès (1999) *Geostatistics: Modeling Spatial Uncertainty.* John Wiley & Sons, New York.
14. G. Ciach, M. L. Morrissey, and W. F. Krajewski (2000) Conditional bias in radar rainfall estimation. *J. Appl. Meteorol.*, 39(11):1941–1946, .
15. J. Creutin and C. Obled (1982) Objective analyses and mapping techniques for rainfall fields: An objective comparison. *Water Resour. Res.*, 18:413–431.
16. R. Ferretti, T. Paolucci, W. Zheng, G. Visconti, and P. Bonelli (2000) Analyses of the precipitation pattern in the Alpine region using different cumulus convection parameterizations. *J. Appl. Meteorol.*, 39:182–200.
17. C. Frei and E. Häller (2001) Mesoscale precipitation analysis from MAP SOP rain-gauge data. *MAP Newsletter*, 15:257–260.
18. Ch. Frei, J. H. Christensen, M. Déqué, D. Jacob, and P. L. Vidale (2003) Daily precipitation statistics in regional climate models: Evaluation and intercomparison for the European Alps. *J. Geophys. Res.*, 108(D3):4124–4142.
19. P. Goovaerts (1999) Performance comparison of geostatistical algorithms for incorporating elevation into the mapping of precipitation. http://www.geocomputation.org/1999/023/gc_023.htm
20. L. Grasso (2000) The differentiation between grid spacing and resolution and their application to numerical modeling. *Bull. Amer. Meteor. Soc.*, 81(3):579–580.
21. B. Ivancan-Picek, D. Glasnovic, and V. Jurcec (2003) Analysis and Aladin prediction of a heavy precipitation event on the Eastern side of the Alps during MAP IOP 5. *Meteorol. Z.*, 12(2):103–112.
22. R. Mladek, J. Barckicke, P. Binder, P. Bougeault, N. Brzovic, C. Frei, J. F. Geleyn, J. Hoffman, W. Ott, T. Paccagnella, P. Patruno, P. Pottier, and A. Rossa (2000) Intercomparison and Evaluation of Precipitation Forecasts for MAP Seasons 1995 and 1996. *Meteorol. Atmos. Phys.*, 72:111–129.
23. A. H. Murphy and R. L. Winkler (1987) A general framework for forecast verification. *Monthly Wea. Rev.*, 115:1330–1338.
24. E. J. Pebesma and C. G. Wesseling (1998) Gstat: A program for geostatistical modelling, prediction and simulation. *Comp. Geosc.*, 24(1):17–31.
25. F. Rubel and M. Hantel (1999) Correction of daily rain gauge measurements in the Baltic Sea drainage basin. *Nordic Hydrol.*, 30:191–208.
26. C. Schär (2001) Alpine numerical weather prediction 2000–2020: A look back to the future. *MAP Newsletter*, 14:6–12.
27. W. Schöner, I. Auer, R. Böhm, and S. Thaler (2003) Qualitätskontrolle und statistische Eigenschaften ausgewählter Klimaparameter (Temperatur, Niederschlag, Schneehöhe) auf Tageswertbasis im Hinblick auf Extremwertanalysen. Report by ZAMG, Vienna. available from W. Schöner, ZAMG, Vienna, Austria.
28. B. Sevruk (1997) Regional dependency of precipitation–altitude relationship in the Swiss Alps. *Climatic Change*, 36:355–369.
29. B. Tustison, D. Harris, and E. Foufoula-Georgiou (2001) Scale issues in verification of precipitation forecasts. *J. Geophys. Res.*, 106(D11):11775–11784.
30. E. Waymire, V. K. Gupta, and I. Rodriguez-Iturbe (1984) A spectral theory of rainfall intensity at the meso-β scale. *Water Resour. Res.*, 20:1453–1465.

31. D. S. Wilks (1995) *Statistical Methods in the Atmospheric Sciences*, volume 58 of *International Geophysics Series.* Academic Press, San Diego.
32. C. Wilson (2001) Review of current methods and tools for verification of numerical forecasts of precipitation. COST717–Working Group Report WDF_02_200109_1, available at http://www.smhi.se/cost717/
33. WMO, Commission for Instruments and Methods of Observation (2001) Expert meeting on rainfall intensity measurements. Technical report, World Metorological Organization.
34. C. B. Young, B. R. Nelson, A. A. Bradley, J. A. Smith, C. D. Peters-Lidard, A. Kruger, and M. L. Baeck (1999) An evaluation of NEXRAD precipitation estimates in complex terrain. *J. Geophys. Res.*, 104(D16):19691–19703.

Robust Spatial Correlation Analysis of the ETEX-1 Tracer Data

Syed Shibli[1] and Gregoire Dubois[2]

[1] Landmark Eame Ltd, Aberdeen, Scotland, UK
syed.shibli@googlemail.com
[2] Radioactivity Environmental Monitoring, Institute for the Environment and Sustainability, Joint Research Centre, European Commission, Ispra, Italy

1 Introduction

At 16:00 UTC on October 23, 1994, 340 kg of perfluoromethylcyclohexane (PMCH) were released into the air from Monterfil in Brittany, France. Air samples were collected at 168 stations in 17 European countries for a period of 90 hours from the start of the release. The European Tracer Experiment (ETEX) was initiated with the aim of collecting data for validating long range transport and dispersion models used for emergency response applications [4, 13]. Another release was made a month later under different meteorological conditions.

Although the data have been used in numerous mechanistic atmospheric dispersion studies, only previously was a geostatistical analysis performed in order to provide some basis for a spatial interpolation [5]. Such an analysis applied fractional Brownian motion models in order to summarise the spatial correlation structure in terms of the power exponent of the variogram, which is directly related to the fractal dimension.

Because of the distressing nature of the data set, which is highly skewed and required a logarithmic transformation in the Dubois et al study, this paper attempts to apply more robust variography on the raw data in order to extract some order out of the chaos. We use the term "robust" in this paper specifically to refer to the stability of variogram in the presence of a strong direct proportional effect, which characterises the ETEX-1 dataset.

2 Analysis of Non-stationary Data

Non-stationary data, including skewed data showing a strong proportional effect, poses significant challenges for spatial correlation analysis and subsequent interpolation. Often it is difficult to determine, from looking at the sample variogram alone, whether what appears to be a power law variogram, representing stationary data with an infinite capacity for dispersion, is not just the manifestation of a quadratic growth due to a non-constant mean.

In theory if we assume that the regionalised variable is made up of two components:

$$Z(s) = \mu(s) + Y(s) \tag{1}$$

where $Y(s)$ is intrisically stationary with zero mean, and $E\{Z(s)\} = \mu(s)$, i.e. the mean varies with location, then we can define the variogram of $Z(s)$ as

$$2\gamma_Z = Var\{Z(s+h) - Z(s)\} \tag{2}$$

or, substituting (1) into (2)

$$\begin{aligned} 2\gamma_Z(h) &= Var\{[Y(s+h) - Y(s)] + [\mu(s+h) - \mu(s)]\} \\ &= Var\{Y(s+h) - Y(s)\} \end{aligned} \tag{3}$$

equation (3) shows that for $Y(s)$ that is intrisically stationary, then the variogram of $Y(s)$ should be equivalent to the variogram of $Z(s)$. However, in practice, the *sample* variogram estimator is typically based on

$$2\gamma_Z(h) = E\{[Z(s+h) - Z(s)]^2\} \tag{4}$$

which would only be valid for data with constant mean. Again, substituting (1) into (4) would give the following expression for the variogram

$$\begin{aligned} 2\gamma_Z(h) = E\{(Y(s+h) - Y(s))^2\} + \{[(Y)(s+h) - Y(s))(\mu(s+h) - \mu(s))]\} \\ +E\{(\mu(s+h) - \mu(s))^2\} \end{aligned} \tag{5}$$

Since $Y(s)$ is intrinsically stationary, the second term on the right hand side of (5) should be zero, so we are left with the expression

$$2\gamma_Z(h) = 2\gamma_Y(h) + [\mu(s+h) - \mu(s)]^2 \tag{6}$$

equation (6) shows that if the mean is constant everywhere, $\mu(s+h) = \mu(s)$, so the variogram for $Z(s)$ should be equivalent to that for $Y(s)$ and (4) would thus apply. However, if the mean is not constant then we will derive a variogram estimator that will exhibit a quadratic growth with h, which would make the estimator invalid.

One method of incorporating a non-constant mean is to estimate a mean surface and work with residuals assumed to be intrisically stationary, e.g. median polish kriging [1]. Nevertheless, in practice, data showing a strong proportional effect (heteroscedasticity) might still require a rescaling of the variogram. This is because the dispersion of the data now depends on its local mean.

Since heteroscedasticity is commonly associated with highly skewed data, one common approach is to transform the data to a logarithmic scale and perform variography and kriging in log space before back transforming, e.g. lognormal kriging [2, 8]. Another approach would be to use alternative measures such as the relative variogram in order to account for the proportional effect via some form of rescaling of the sample variogram values.

Both techniques of transformation and use of the relative variogram are related, as demonstrated by Cressie [1]. If we assume that we have spatial regions$\{D_j; j = 1, \ldots, n\}$ within which the regionalised variable $Y(s)$ is intrinsically stationary with mean μ_j and variogram $2\gamma_Z^j(h)$ for each region $j = 1, \ldots, n$, then using the δ method of Kendall and Stuart [9] we can apply a transformation for the variable $Z(s)$ as follows:

$$Y^j(s) = g[Z^j(s)]; s \in D_j \tag{7}$$

where $g(\cdot)$ is sufficiently smooth to possess at least two continuous derivatives. Kendall and Stuart applied a Taylor series expansion about $E(Z(s))$, i.e.

$$Y^j(s) = g(\mu_j) + g'(\mu_j)[Z^j(s) - \mu_j] + g''(\mu_j)[Z^j(s) - \mu_j]^2/2! + \ldots; s \in D_j \tag{8}$$

It follows from (7) that if we were to take the increments for $Y^j(s)$ then

$$Y^j(s+h) - Y^j(s) = g'(\mu_j) + g(\mu_j)[Z^j(s+h) - \mu_j] + \ldots; s \in D_j \tag{9}$$

We can then define the variogram for $Y^j(s)$ by applying the variance operator on both sides of (9) to derive

$$2\gamma_Y^j(h) = (g'(\mu_j))^2 2\gamma_Z^j(h); j = 1, \ldots, k \tag{10}$$

(10) is similar to the general form of the local relative variogram, defined by various authors [7, 8] in the following manner:

$$2\gamma_{RY}(h) = \frac{2\gamma_Z^j(h)}{\mu_j^n} \tag{11}$$

which is independent of j and where n is typically 2. Equating (10) and (11) we note that when $g(\mu_j) = 1/\mu_j$ the process $Y(s)$ is approximately intrinsically stationary, i.e.

$$g(x) = \log(x) \tag{12}$$

This result shows that the relative variogram can be used as an alternative to log transformation, in order reduce the influence of a proportional effect on the traditional sample variogram. Srivastava and Parker [12] present another *deterministic* form of the relative variogram, namely the pairwise relative variogram described as follows:

$$2\hat{\gamma}_{PR}(h) = \frac{1}{N_h} \sum \left[\frac{z(s+h) - z(s)}{z(s+h) + z(s)} \right]^2 \tag{13}$$

The above takes each squared difference between pairs of sample values and divides it by the square of their average, hence the term "pairwise."

Srivastava [11] also presents non-ergodic versions of the covariance and correlogram functions to accomodate the proportional effect. Starting with the definition of the covariance, i.e.

$$C(h) = E\{Z(s)\}E\{Z(s+h)\} - \mu(s+h) \tag{14}$$

Srivastava defines the following estimator for (14)

$$\hat{C}(h) = \frac{1}{N(h)} \sum z(s+h)z(s) - \overline{z}(s)\overline{z}(s+h) \tag{15}$$

with the caveat that, although $\hat{\gamma}(h) \neq \hat{C}(0) - \hat{C}(h)$, the differences should be sufficiently small that one can apply $\hat{\gamma}(h)$ in practice. Curriero et al. [3] contends that the "head" and "tail" values in (15) are meaningless for the omnidirectional direction since the common practice is to count the location twice. Hence the re-scaling of the variance is implicitly taken into account by incorporation of the mean for all the data contributing to the lag.

A common problem with using (15) to derive an approximate sample variogram is that the first few lags can result in negative values, since the number of data used to derive the sample variance is typically greater than the number of data used to derive the sample covariance itself. Nevertheless, such a difficulty does not exist if one scales each covariance value by the product of the standard deviations for the head and tail values, thus giving the following definition for the correlogram:

$$\hat{\rho}(h) = \frac{\hat{C}(h)}{\hat{\sigma}(s)\hat{\sigma}(s+h)} \tag{16}$$

By explicitly accounting for the possibility that some lags contain more variable values than others, the correlogram is likely to suffer least from the combination of heteroscedasticity and clustering [?].

Yet another variation of (11) presented by Isaaks and Srivastava [7] is the general relative variogram, defined as follows:

$$\hat{\gamma}_{GR}(h) = \frac{\hat{\gamma}(h)}{\frac{1}{2N_h} \sum z(s) + z(s+h)} \tag{17}$$

where $\hat{\gamma}(h)$ is the traditional method of moments variogram estimator. Equation (17) circumvents the potential problem of having too few pairs in the local disjoint regions $j = 1, \ldots, n$, which is required in order to estimate the local relative variogram defined by (11).

3 Data Description

The ETEX-1 data consists of 155 raw concentration measurements of PMCH (in units $\mathrm{ng/m^3}$) from the first ETEX release recorded at 26 different times, from 15 hours to 90 hours in increments of three hours, to give a total of 4,030 measurements. Thirteen stations out of the original list of 168 were discarded a priori; nine due to the absence of any reliable measurements and four being dismissed as outliers [5]. PMCH is an inert gas so is insensitive to wet scavenging or uptake at the surface and thus is only affected by advection and diffusion.

The time $t = 15$ hours was chosen as the start time since by this time all of the tracer had been released to the atmosphere. A preliminary analysis of the data indicated that a maximum of 95% of the stations did not record any presence of the tracer at $t = 15$ hours and a minimum of 69% was found at $t = 45$ hours. Due to the high positive skewness of the data, in the original paper a logarithmic transformation was applied to minimise the effect of the skewness on the variogram.

4 Univariate Statistical Analysis

Figure 1 shows the location map of the monitoring stations, which covers part of Western Europe. The maximum diagonal distance for the analysed area is some 3,093 km, and a quick analysis of the geographical locations indicates that there is no preferential sampling strategy and that their spatial distribuion appears to be random (Clark and Evans's index=1.032). No declustering was therefore performed on the data.

When plot on a time axis, we observe the mean of the concentrations to increase between $t = 15$ hours and $t = 45$ hours before decreasing to zero thereafter right up to $t > 80$ hours (Fig. 2) due to the large proportion of zero measurements. The variance tracks a similar behaviour; the decrease in the first and second order moments beyond $t = 60$ hours could possibly be due to the cloud dispersing at a faster rate and splitting into more distinct parts.

Skewness and kurtosis (Fig. 3) are quite constant in time from $t = 15$ hours to about $t = 81$ hours, indicating no general changes in shape of the histograms (highly positively skewed to the right) for almost all time slices.

Figure 4 shows a direct proportional effect in time for the mean concentration versus variance. The correlation coefficient between the mean and variance is ca. 83%, which is high. Moving window statistics were also derived for the data for $t = 45$ hours, and Fig. 5 shows that the linear relationship between the mean concentration and variance is also present both in the east–west and north–south directions.

Finally, Figs. 6–8 present the probability plots of the surface concentrations for times from $t = 15$ hours to $t = 75$ hours. The skewness is highly evident from these plots, with zero values making up a large proportion of the recorded data. At times greater than 63 hours, the proportion of zeros increases.

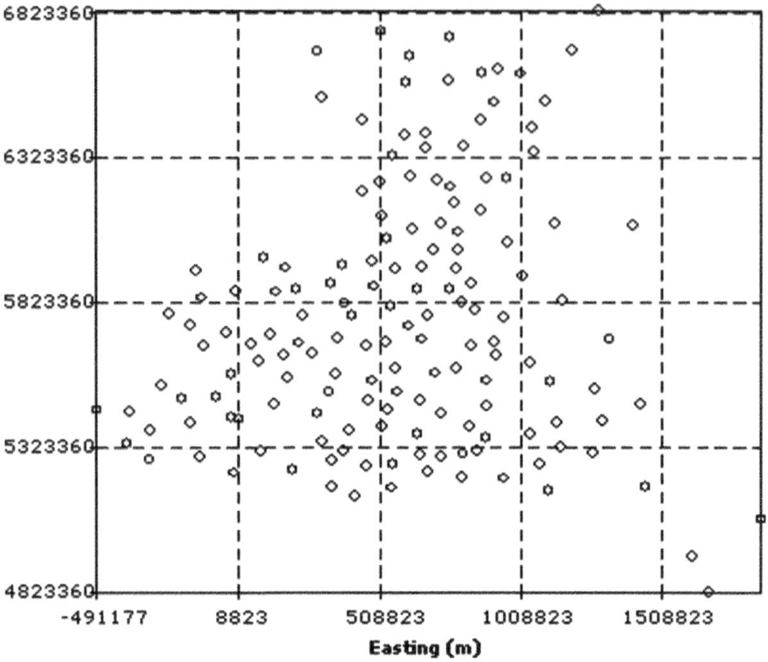

Fig. 1. Location of ETEX-1 monitoring stations

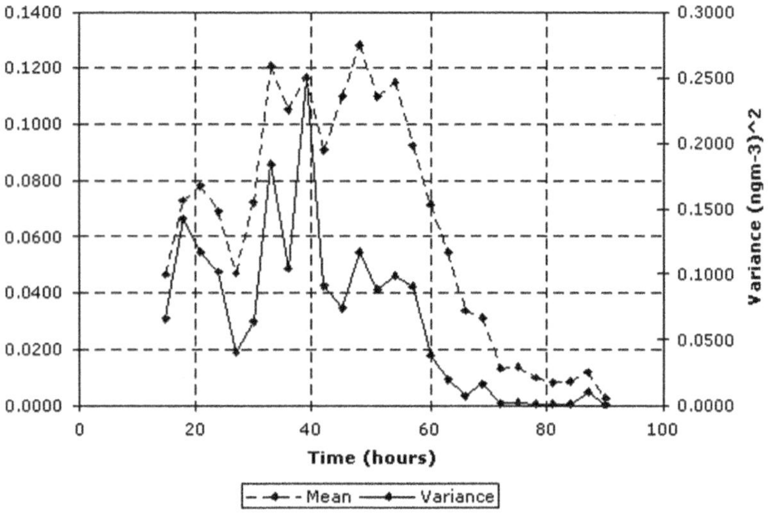

Fig. 2. Tracer concentration mean and variance over time

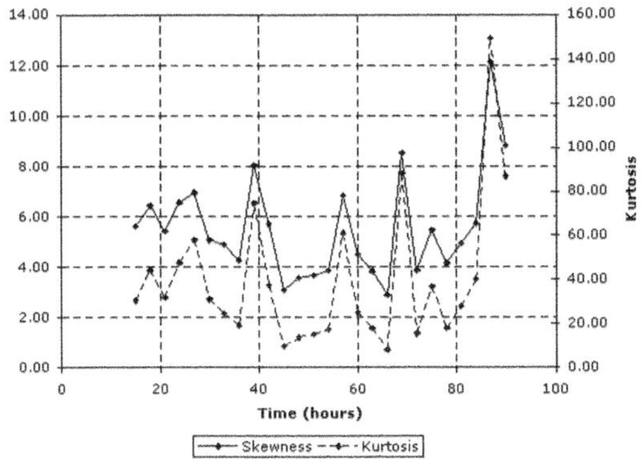

Fig. 3. Skewness and kurtosis over time

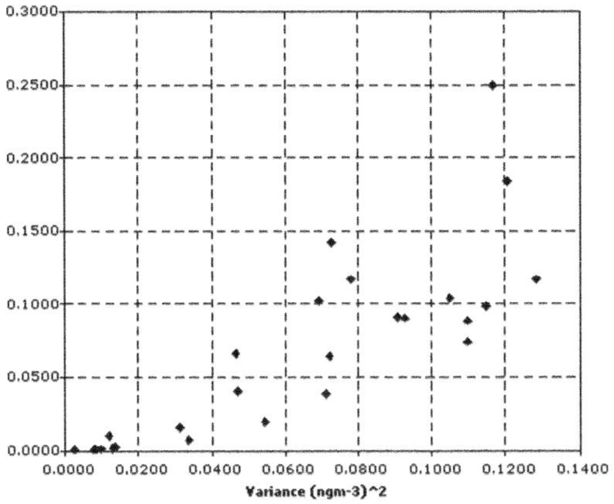

Fig. 4. Tracer concentration mean versus variance for all times

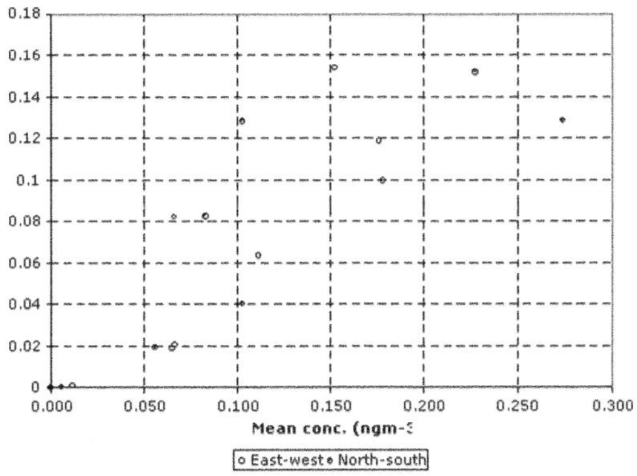

Fig. 5. Moving window statistics at $t = 45$ hours

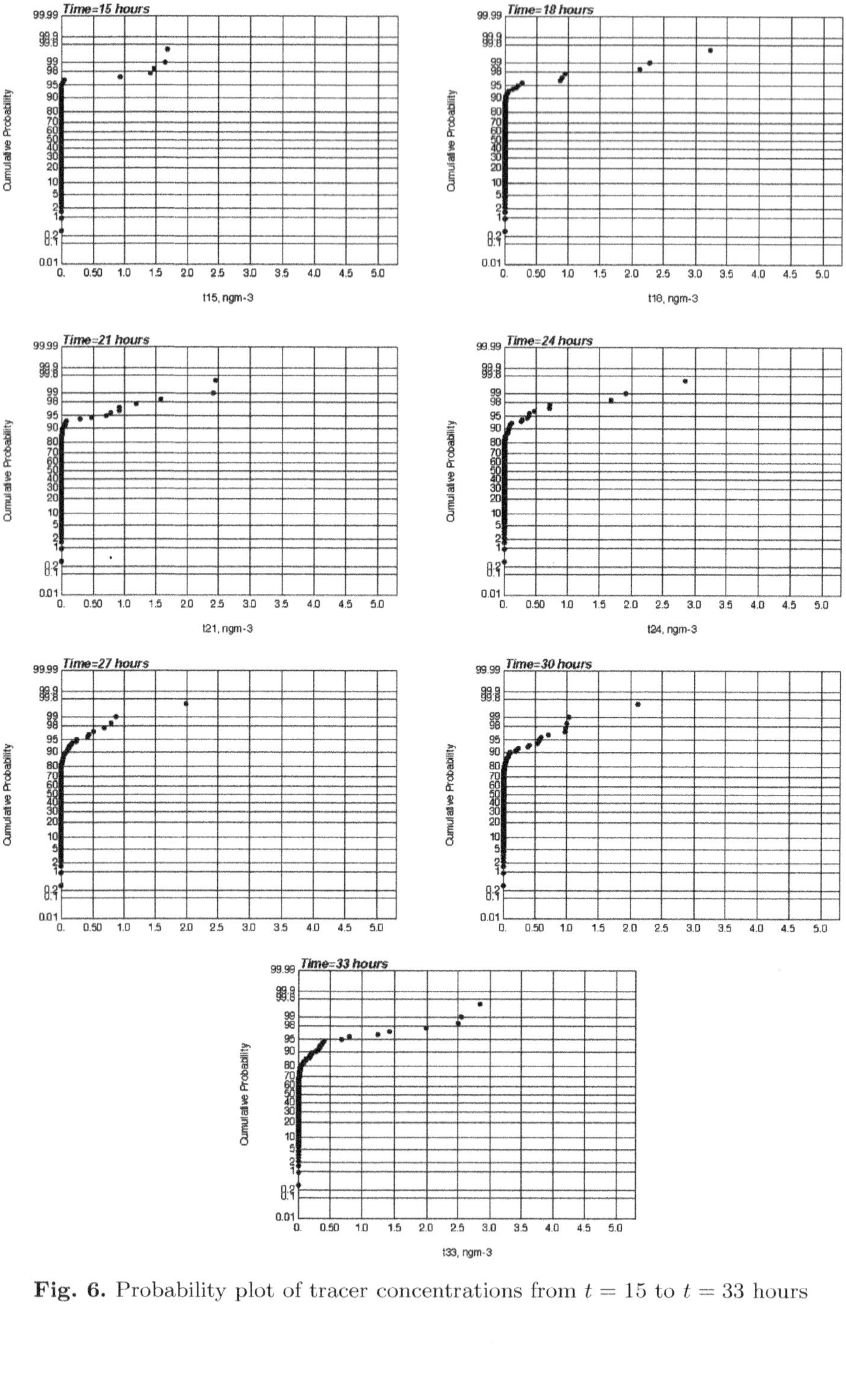

Fig. 6. Probability plot of tracer concentrations from $t = 15$ to $t = 33$ hours

5 Spatial Correlation Analysis

An analysis of the spatial correlation structure of the surface concentration data is a prerequisite for spatial interpolation, whether for estimation (e.g. kriging) or simulation (e.g. sequential Gaussian simulation). This phase of the analysis quantifies, via a traditional measure such as the sample variogram, the spatial dependency between data separated some lag distance, h, in space or time, described by (2) or (4).

In ordinary kriging, the estimated value $\hat{z}(s)$ is given by

$$\hat{z}(s_0) = \sum_{i=1}^{n} \lambda_i z(s_i) \tag{18}$$

where λ_i are weights assigned to the observed data $z(s_i)$ that will determine their role in defining the value taken by the variable at unsampled location s_0. The main interest in applying geostatistical techniques is that these weights are computed from a model of the spatial correlation of the analysed phenomena. Hence, unlike other interpolators (for an overview of interpolation methods refer to Lam [10]), geostatistics takes the spatial structure of the variable explicitly into account.

A useful opening move in any geostatistical study is to derive the omnidirectional semivariogram of the variable under study to determine the degree of spatial correlation, often called a "structural analysis". As noted earlier, the heteroscedasticity observed in the data can makes the inference of a range and sill very difficult using conventional measures such as the sample variogram. Although by itself the shape of the semivariogram may not be affected by the heteroscedasticity if the mean of the sample values is roughly the same for all lags, the oridinary kriging variance, however, is dependent on the magnitude of the variogram.

For the sake of brevity, the analyses will now be presented based on the data for $t = 45$ hours. Figure 9 shows six different spatial correlation measures for this time slice, calculated using an angular tolerance of 90 degrees (omnidirectional). Twelve lags were calculated at a lag increment of roughly 100 km, with a lag tolerance of about 50 km. This ensured that there were at least 30 pairs for each lag distance.

The functions shown in Fig. 9 also include the sample general relative variogram (17), pairwise relative variogram (13), non-ergodic covariance (14), non-ergodic correlogram (16), and semimadogram (a measure of the average mean absolute difference), defined by the following:

$$2\hat{\gamma}_M(h) = \frac{1}{N_h} \sum |z(s+h) - z(s)| \tag{19}$$

It can be noted that the both the madogram and variogram provide a poor measure of the spatial correlation for this highly skewed data set, resulting in an erratic sample variogram. Both measures rely only on the mean difference

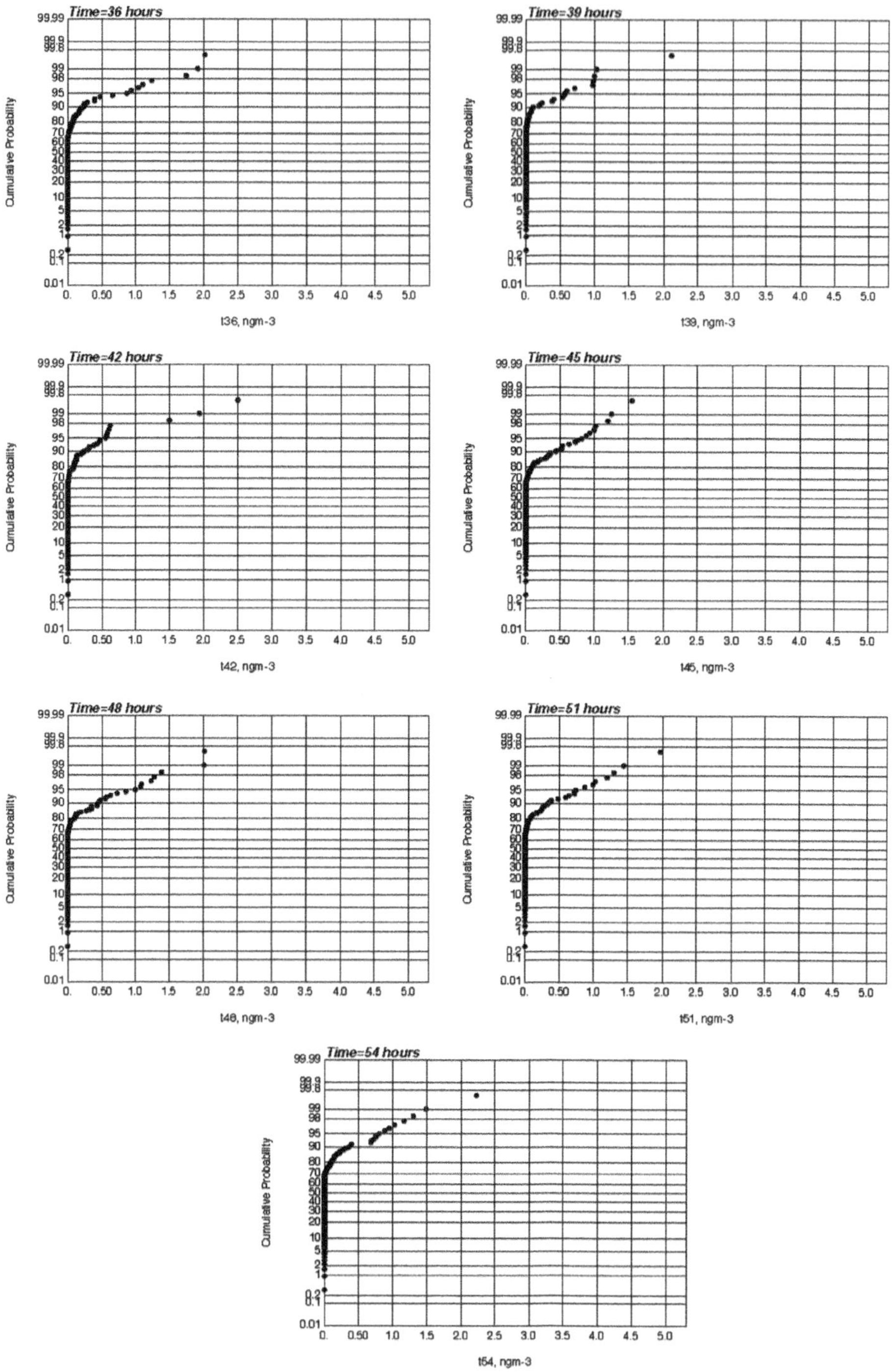

Fig. 7. Probability plot of tracer concentrations from $t = 36$ to $t = 54$ hours

(squared in the case of the variogram) between two data points located h m apart, so no rescaling of the variogram is performed commensurate with the proportional effect.

The madogram also forms the basis for other well known "robust" variography techniques, e.g. that of Cressie and Hawkins [2] using the fourth power of the square root of the madogram; and that of Genton [6] based on the k-th quantile of the madogram. However these techniques rely on stability of the variogram based on deviations from a primarily Gaussian distribution, and are not expected to fare very well in the presence of a strong proportional effect.

From Fig. 9, for the other four measures which scale as some function of the mean value, some semblance of *structure* can be inferred, giving an omnidirectional range for the tracer concentration values at around 40,000 m. The pairwise relative variogram results in a usable albeit smooth model because values at each pair are rescaled by the mean of the values contributing to that pair. The correlogram appears to be less erratic than the covariance, due to the advantage of having the values standardised by the data variances. Both the correlogram and the pairwise relative variogram result in smoother spatial correlation structures compared to the general relative variogram and covariance.

The above results confirm that some sort of spatial dependence exists for the concentration cloud at $t = 45$ hours, which was not evident from the traditional variogram measure. Figure 10 shows the directional pairwise relative variogram for four principal directions 0 deg, 45 deg, 90 deg, and 135 deg based on an angular tolerance of 22.5 deg. Little or no anisotropy is evident; for the direction 135 deg a somewhat shorter range of some 300 km is observed, and this appears to be the direction of minimum continuity. The range in the direction of maximum continuity appears to be between 400 and 500 km.

For convenience, and as a prerequisite for spatial interpolation, we will model the spatial variability by assuming a power law model for the sample variogram. This is performed by curve fitting a log–log plot of the pairwise relative variogram values versus lag distance (refer Fig. 11) and inferring the slope of the fit.

The variance of increments for random fractional Brownian motion (fBm) models satisfying a distribution with fractal geometry can be written as:

$$2\hat{\gamma}(h) = V_H(h)^{2H} \tag{20}$$

where H is the intermittency or Hurst exponent and is related to the fractal dimension $D = 2 - H$. The intermittency exponent falls between zero and one; the special case $H = 0.5$ characterising normal Brownian motion. For $H < 0.5$, the phenomena is anti-persistent (more zero crossings) and less continuous; for $H > 0.5$ the phenomena is persistent, or values tend to be clustered together and high values are more likely to be followed by high values as well.

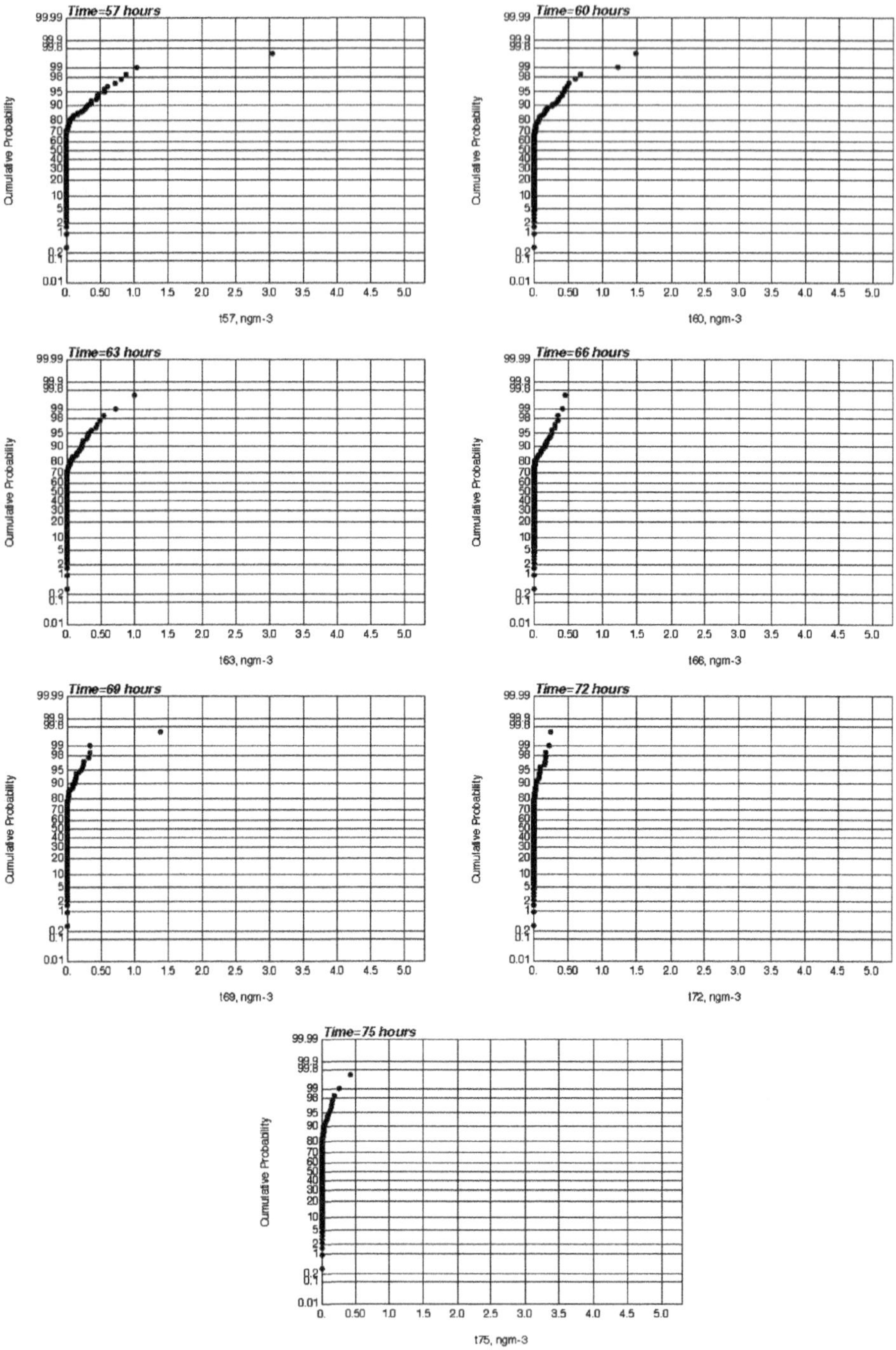

Fig. 8. Probability plot of tracer concentrations from $t = 57$ to $t = 75$ hours

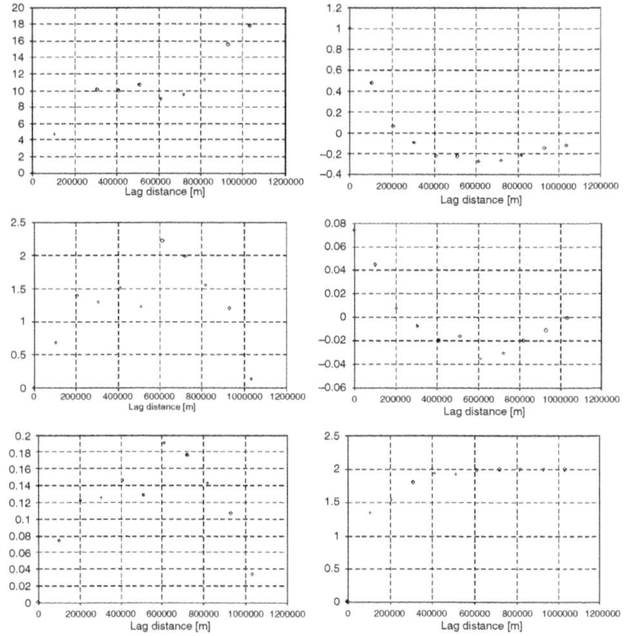

Fig. 9. Spatial correlation measures at $t = 45$ hours (omnidirectional)

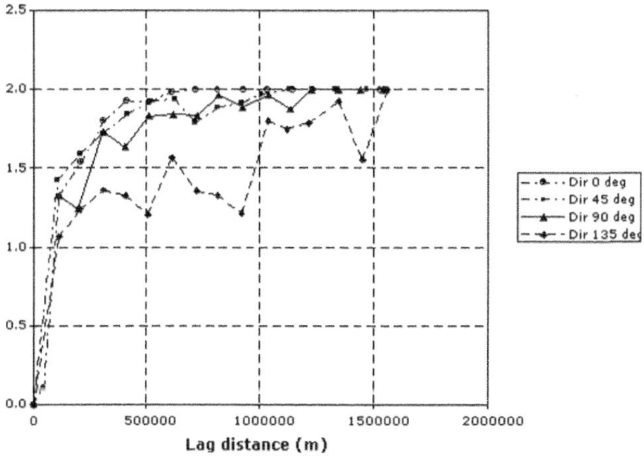

Fig. 10. Directional pairwise relative variograms at $t = 45$ hours

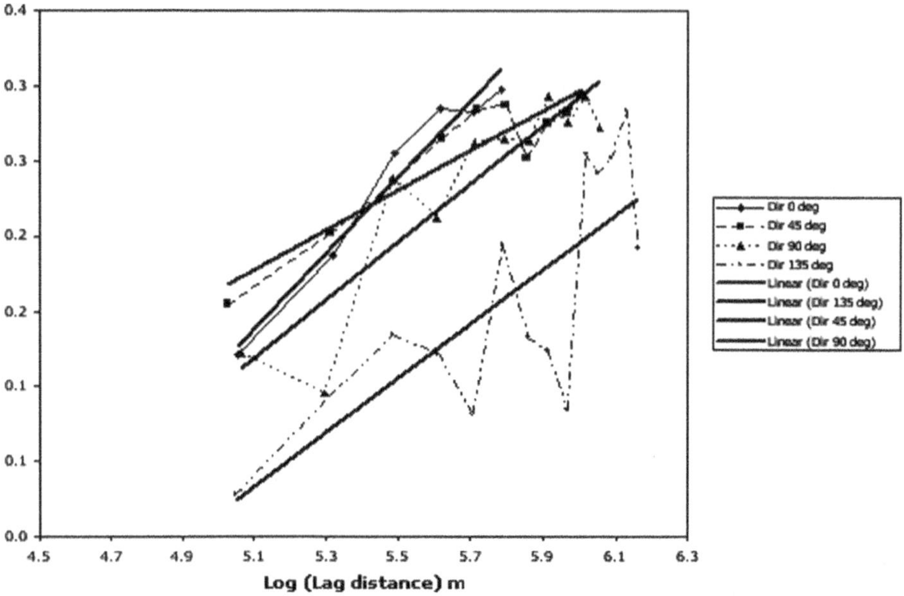

Fig. 11. Log–log model fit for directional variograms

The four directional variograms in Fig. 10 give the following fractal dimensions:

direction	fractal dimension
0 deg ± 22.5 deg	1.87
45 deg ± 22.5 deg	1.93
90 deg ± 22.5 deg	1.90
135 deg ± 22.5 deg	1.91

The dimensions are characteristically high, although within the same range of values reported by Dubois et al. [5], who based their results on log–log plots of variograms based on log transformed variables. These high dimensions translate to very low intermittency exponents, corresponding to anti-persistence, or less continuous phenomena. Such behaviour is also reflected in the shape of the relative variograms themselves.

6 Discussion

For the ETEX-1 data, the traditional variogram estimator performs poorly in the presence of a strong proportional effect, showing erratic behaviour at all lags. Use of the alternative estimators which re-scale the variogram value by some function of the mean allows us to infer large scale structure from the highly skewed data. This obviates the need to perform a logarithmic transform in order to temper the impact of such skewness. The next step in the analysis would be to perform a spatial interpolation of the concentrations based on the inferred structural range and curvature. Use of power law (fBm) models is one alternative.

References

1. Cressie, N. (1993) *Statistics for Spatial Data*, Revised edition. John Wiley & Sons.
2. Cressie, N.A.C and Hawkins, D.M. (1980). Robust Estimation of the Variogram: I. Mathematical Geology, 12(2):115–125.
3. Curriero, F., Hohn, M., Liebold, A., and Lele, S. (2002). A statistical evaluation of non-ergodic variogram estimators. Environmental and Ecological Statistics, 9(1):89–110.
4. Girardi, F., Graziani, G., van Veltzen, D., Galmarini, S., Mosca, S., Bianconi, R., Bellasio, R. and Klug, W. (eds.) (1998). "The ETEX Project". EUR Report 18143 EN. Office for Official Publications of the European Communities, Luxembourg.
5. Dubois, G., Galmarini, S., and Saisana, M. (2005). Geostatistical Investigation of ETEX-1: Structural Analysis. Atmospheric Environment, 39: 1683–1693.
6. Genton, M. (1998). Highly Robust Variogram Estimation. Mathematical Geology, 30(2):213–221.
7. Isaaks, E. and Srivastava, R.M. (1989). An Introduction to Applied Geostatistics, Oxford University Press, Oxford.
8. Journel, A.G. and Huijbregts, C.J. (1978). Mining Geostatistics. Academic Press, New York.
9. Kendall, M.G., and Stuart, A. (1969). The advanced theory of statistics, Vol. 1, 3rd ed. Griffin, London.
10. Lam, N.S. (1983). Spatial interpolation methods: a review. The American Cartographer, 10(2):129–149.
11. Srivastava, R.M. (1987). A Non-ergodic framework for variograms and covariance functions. M.Sc. Thesis, Stanford University.
12. Srivastava, R.M. and Parker, H.M. (1988). Robust Measures of Spatial Continuity, In: Geostatistics Volume 1, Proceedings of the Third International Geostatistics Congress, Kluwer Academic Publishers.
13. van Dop, H. and Nodop, K. (eds). (1998). ETEX: A European Tracer Experiment, Atmospheric Environment 32:4089–4378.

Fuzzy Model of Soil Polygons for Managing the Imprecision

Rangsima Sunila and Pekka Horttanainen

Department of Surveying, Institute of Cartography and Geoinformatics, Helsinki University of Technology (HUT), Espoo, Finland
rangsima.sunila@hut.fi

1 Introduction

It is impossible to create a perfect representation of the world in a GIS database since all GIS data are subject to uncertainty [5]. There is no perfect information that contains 100% accurate presentation in a GIS database. The impression of certainty usually conveyed by GIS is at odds with the uncertain nature of geographic information, a contradiction that has been acknowledged as an important research topic for nearly two decades [4]. Incorrectness in measurement or errors in observations due to rich information bring about imperfection in geographic information. The information is taken and used as if it were accurate or believed to be true. In fact, the reliability of the information is not yet considered in terms of level of accuracy or uncertainty. If the geographic information is looked over carefully, it contains vagueness, imprecision and inaccuracy particularly when it presents an invisible object like soil. Soil polygon boundaries are defined based on accurate field observation and compared with human interpretation and soil types are classed according to geologists' prowess or expertise. For many reasons, one can say that soil map is one of the most imprecise maps in the world.

1.1 The Background of the Research

In Finland, mapped areas are big and field observations are relatively sparse. Manual interpretation is therefore the only feasible alternative in creating soil maps. Samples of soil are taken randomly by soil mapping surveyors for soil type classification and some of these samples may be taken back to the laboratory for detailed tests in case of inadequacies (Fig. 1).

For defining soil polygon boundaries, geologists use aerial photos, geologic maps, topographic maps in scale 1:20,000 together with knowledge based of geomorphology. Nevertheless, there are no specific rules to define the imprecision of these boundaries and neither imprecision in data nor expert knowledge

Fig. 1. Soil mapping fieldtrip in Porvoo, Finland

is recorded. As a result, imprecision in soil data is carried along the survey process and it continues to the stage when the data is entered to soil database then the final result comes out in the form of printed map that certainly contains uncertainty information.

The problem to be solved is to try to collect and describe the knowledge and experience of geologists. To cope with this and discussion about uncertainty in geographic databases, fuzzy logic will be used as a tool to describe and solve the problem of modeling imprecise objects like soil polygons.

1.2 The Goal

The core idea in this research is to apply fuzzy modeling to the management of expert knowledge in soil mapping. Fuzzy soil maps are then used in map overlay type analysis [11]. The goal is not to construct a fuzzy model of soil layers but a fuzzy soil map presenting non-crisp soil polygon boundaries. The map will be created in certain scale for a certain purpose and we believe that fuzzy soil layer with imprecision is better input to the analysis than artificial crisp polygon map with no information about the uncertainty of the boundaries.

2 Literature Survey

Recently, there have been many researches that were related to fuzzy concepts. Brown [2] carried out a research in classification and boundary vagueness in mapping presettlement forest types. In his research, he conducted a model to test the role of classification ambiguity in affecting boundary vagueness using fuzzy concepts and Kriging. He explained methods of determining species memberships and interpolating membership values. Stefanakis et al. [10] conducted a research on incorpolating fuzzy set methodologies in a Database Management System (DBMS) repository for the application domain of GIS. They considered that fuzzy set methodologies seemed to be instrumental in the design of efficient tools to support the spatial decision-making process. The results showed that Zadeh's fuzzy concepts and fuzzy set theory [12] might be adopted for the representation and analysis of geological data. Jiang et al. [6] proposed the application of fuzzy measures and argued that the standardized factors of multi-criteria evaluation belong to a general class of fuzzy measures and the more specific instance of fuzzy membership. Developing of classification algorithms for using auxiliary information in fuzzification and fuzzy set operations to reduce uncertainty in classification process was researched in 2000 by Oberthür et al. [9]. The research was conducted in order to study how to define fuzzy membership functions (FMF) and reduce classification uncertainty hedge operators. Zhu et al. [13] introduced soil mapping using GIS, expert knowledge, and fuzzy logic. The scheme consisted of three major components: a model employing a similarity representation of soils, a set of inference techniques for deriving similarity representation and use of the similarity repreentation. Basically they invented an automated soil inference under fuzzy logic. To produce a raster soil database for the study areas, the knowledge base and the spatial data in the GIS database were combined under the fuzzy inference engine. The output was the comparison of soil series referred from Soil Land Inference Model (SoLIM) and derived from the soil map against the field observations for the study area and it showed that SoLIM has higher correctness. The derivation of the fuzzy spatial extent was developed by Cheng et al. [3]. Three fuzzy object models and the data extracted from field observation were introduced and modeled. Software such as FUZZEKS [1], FuzME Version 3.0 [8] and ASIS [7] were programmed to deal with fuzzy logic and data analysis.

3 Pilot Studies

At this stage of the project, expert knowledge is being collected. The research team is currently trying to document a rule-based model of soil polygon boundaries. As the geographic environment is varied depending on regions and geomorphy, each region needs to be differentiated and taken into consideration separately. To construct a fuzzy model, the values of membership functions are required. The membership functions from different regions should present different values, as the behavior of geographic data in each region is distinct. This may require a lot of work in the beginning but once the system is set up, everything will be processed smoothly. When fuzzy membership functions of soil types are defined, fuzzy models of soil polygons will be constructed.

4 Methodology

From the dataset received from Geological Survey of Finland, the data will be read in numerical format as in Fig. 2. These numbers represent different types of soil and the connection with different numbers implies where the polygon borders are. As it was mentioned earlier, to construct fuzzy models, membership functions of classification are needed.

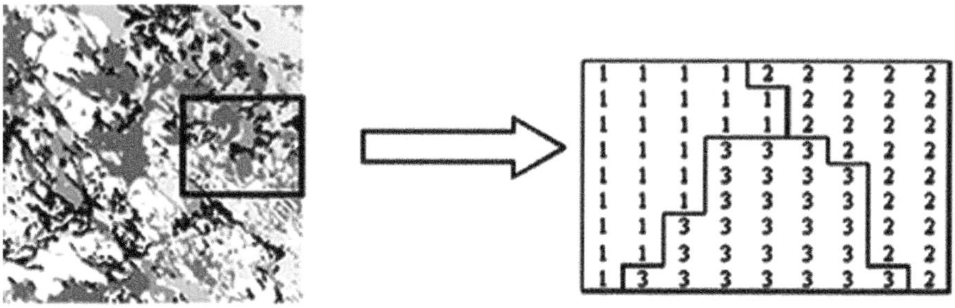

Fig. 2. Example of soil data in numerical format; soil type 1: bedrock, soil type 2: sand, soil type 3: clay

Therefore, Table 1 is a fundamental design for geologists to provide their expert opinions of the level of certainty in soil polygon boundaries. For example, number 0.75 in the table is the value of the pixel's membership function adjacent to the boundary, which means that it is not crisp. The numbers on the diagonal represent the value of the membership function within the soil polygon.

Table 1. Level of certainty of polygon borders between different soil types ranged from 0 to 1

eastern Finland	bedrock	sand	clay
bedrock	0.95	0.92	0.88
sand		0.9	0.75
clay			0.85

From the numbers in the table, logical rules will be constructed in a fuzzy tool. The expected result is the soil data layer that shows imprecision on the soil polygon boundaries. Next phase, the layer of elevation could be added to adjust the values of fuzziness.

5 Expectations

It takes time to start up the system and develop a good connection that will lead to success in the future. Collecting geologists' knowledge and trying to construct the documentation of expert knowledge is the first priority. From the documentation, a rule-based model is created which will lead to the construction of fuzzy models. Currently, there are three expectations.

- Documentation of geological knowledge used in interpretation
- Development of a rule-based model of imprecise soil polygons which could be used for GIS analysis. This is not to create a new soil model for geologists or even for soil mapping but to improve the represetation of soil data layer that shows the imprecision of the classification especially around the boundaries of soil polygons.
- Fuzzy modeling of soil maps to understand uncertainty in geographical information for better uses in spatial analyses

6 Future Plan

For further research, there are still many possibilities to continue studying imprecision in soil polygon boundaries, for instance, implementing Kriging to test out the result of fuzziness. The question may arise here whether Kriging can be used, as the values seem to be from a discrete function. Clearly, Kriging is not going to be used for better classification of soil types, instead, it will be applied together with a fuzzy model to verify the imprecision on soil polygon boundaries to smooth out the result. Sample points could be taken from the real site to study soil type misclassification and these numbers will be used together with fuzzy membership functions for better results.

6.1 An Example

Figure 3 shows an example of fuzzy and Kriging application on soil polygon boundaries.

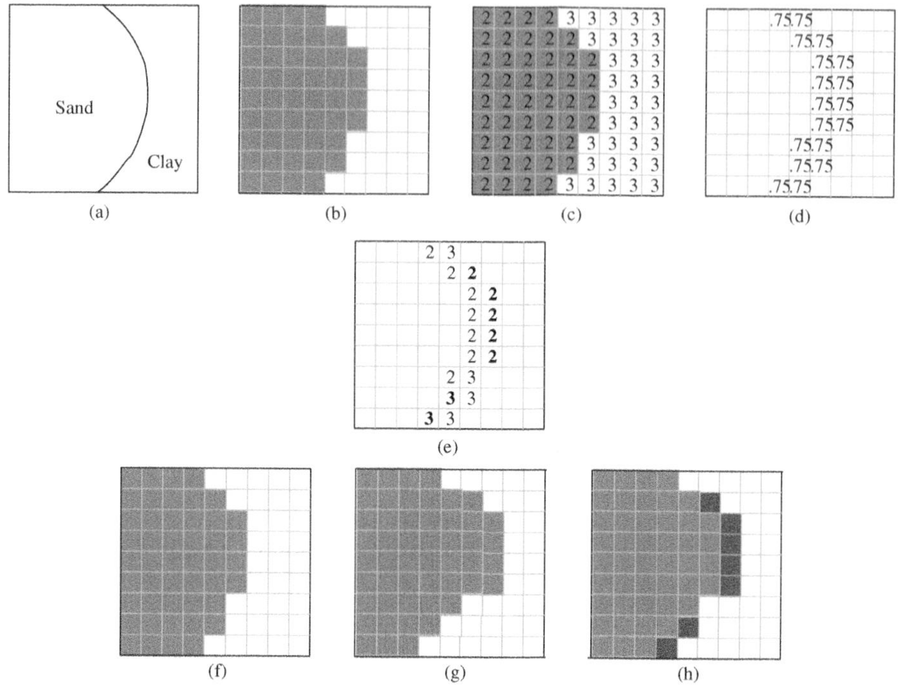

Fig. 3. (**a**) Consider a map that contains only two soil types, sand and clay. (**b**) The soil map data is transferred to a raster format. (**c**) From the raster format the data is coded: sand = 2 and clay = 3. (**d**) Next step, membership functions represent the level of classification certainty on soil polygon boundaries. The result is the data layer that contains imprecise soil polygon boundaries. (**e**) Kriging is used to test out the soil property in each class. In this example, misclassification is discovered and it affects the boundary. Comparison: (**f**) The original soil polygon boundary in raster format. (**g**) The new soil polygon boundary resulted from Kriging. (**h**) The *highlighted* area shows the error or fuzziness along the boundary. The *highlighted* areas are the areas that should be taken into consideration in order to adjust the values of membership functions

7 Difficulties in Using Kriging

Although Kriging is a well-known geostatistical tool, it is not being used in Finnish soil mapping. Finland is big with its size of about 338,000 km^2 so it is difficult to have very high density of soil sampling in a large scale. Moreover, much effort and skill are needed for setting up test areas in every region. This will be too time consuming and expensive, which is an important matter for concern. Besides, the results are believed to be different in each particular area so it is hard to set up a standard value for error measurements. In conclusion, this method has high requirements in order to conduct the research.

However, it is hoped that developing membership functions in fuzzy models together with Kriging would give better solutions in studying imprecise information in soil databases especially in some specific areas that are worth for studying in depth for better spatial analysis.

8 Conclusion

Soil polygon boundaries are not crisp in reality. Moreover, soil maps contain a large amount of uncertainty and imprecision. Therefore a natural way to model vagueness of soil polygons is to include imprecision in their boundaries. One way to do this is to develop a fuzzy model for raster data using fuzzy membership functions for each soil layer. This model will be created using expert knowledge and fuzzy logic.

In Finland there are no sufficient metadata to assess the uncertainty of soil polygon boundaries. Thus, the first step of the research is to collect and document expert knowledge about soil mapping. Only then can the rule-based fuzzy model be created. The information of imprecision provided by the fuzzy model will eventually be used in GIS to give an estimation of uncertainty in soil maps.

Acknowledgements

The project is funded by the Ministry of Forestry and Agriculture of Finland and in cooperation with Geological Survey of Finland. This project is an offspring from the military terrain analysis project conducted by the Scientific Board of the Finnish Defence Forces. The authors would like to thank Professor Kirsi Virrantaus for introducing the project and giving her comments and suggestions, Jukka-Pekka Palmu and Maija Haavisto-Hyvärinen for supporting the information, Professor Vesa Niskanen for instruction of fuzzy modeling, Olga Křemenová for her assistance and cooperation and StatGIS 2003 conference committee for giving a chance to present the project.

References

1. Bartels F (1997) FUZZEKS. The Fuzzy Evaluation and Kriging System. http://www.fuzzeks.de
2. Brown D (1998) Classification and boundary vagueness in mapping presettlement forest types. Int. J. Geogr. Inf. Sci. 12(2):105–129.
3. Cheng T, Molenaar M and Lin H (2001) Formalizing fuzzy objects from uncertain classification results. Int. J. Geogr. Inf. Sci. 15(1):27–42
4. Duckham M (2002) Uncertainty and geographic information: computational and critical convergence. In: Representation in a digital geography, Wiley: New York.
5. Goodchild M (2003) Models for Uncertainty in Area-Class Map Presentation lecture note, Uncertainty in Geographical Information course, Helsinki University of Technology, 2–3 June
6. Jiang H and Eastman J R (2003) Application of fuzzy measures in multi-criteria evaluation in GIS. Int. J. Geogr. Inf. Sci. 14(2):173–184
7. Mazaheri S A, Koppi A J, Mcbratney A B and Constable B (1995) Australian Soil Identification Spreadsheet (ASIS). A program for allocatiing soil profiles to Australian Great Soil Groups (GSG)
8. Minasny B and McBratney AB (2002) FuzME version 3.0, Australian Centre for Precision Agriculture, The University of Sydney, Australia
9. Oberthür T, Dobermann A and Aylhard M (2000) Using auxiliary information to adjust fuzzy membership functions for improved mapping of soil qualities. Int. J. Geo. Inf. Sci. 14(5):431–454
10. Stefanakis E, Vazirgiannis M and Sellis T (1999) Incorporating fuzzy set methodology in a DBMS repository for the application domain of GIS. Int. J. Geogr. Inf. Sci. 13(7):657–675
11. Virrantaus K (2003) Analysis of the Uncertainty and Imprecision of the Source Data Sets for a Military Terrain Analysis Application, Proceedings of the 2nd International Symposium of Spatial Data Quality, Hong Kong, 20–22 March
12. Zadeh L A (1965) Fuzzy sets. Information and Control 8:338–353
13. Zhu A, Hudson B, Burt J, Lubich K and Simonson D (2001) Soil Mapping using GIS, Expert Knowledge, and Fuzzy Logic, Soil Sci. Soc. Am. J. 65:1463–1472

Mapping the Contaminant Legacy of a Coking Plant, The Avenue, Chesterfield, UK

Monica Palaseanu-Lovejoy, Ian Douglas, and Robert Barr

School of Geography, The University of Manchester, Manchester, UK
monica.palaseanu-lovejoy@stud.man.ac.uc

1 Introduction

With 300,000 ha of contaminated land, 1.2% of the Britain land area [13, 14], the UK has a major need for effective environmental risk assessment for land remediation and reclamation [22]. Such a risk assessment is usually based on the characterization of potential site contaminants and analysis of source – pathway – target scenarios [8, 9, 11]. A risk-based contaminant description requires a conceptual model of the site that includes qualitative and quantitative analyses of pollution sources, contaminant pathways and pollutant receptors [4, 22]. This characterization typically has to rely on limited and irregularly distributed point data. In addition, soil and surface material heterogeneity, as well as the quasi-random nature of contamination sources add to the complexity of developing good spatial models of pollutants on old industrial sites [13, 22]. Improved and appropriate geostatistical tools and GIS based analysis can help to overcome some of these problems. This paper tackles the development of such a methodology for a former coking plant by examining the sources and pathways of Polycyclic Aromatic Hydrocarbons (PAHs), as part of an analysis of a wider range of contaminants at the Avenue Coking Works, near Chesterfield, UK (Fig. 1). In the UK, coking works were established alongside the iron and steel plants from the mid 18th century. By the end of the 19th century surplus gas from coking works was sold as town gas, and by 1912 coke ovens were being installed at town gas works [10]. Each works occupied between 0.3 and 200 ha. By 1995, only four of the total of 400 such works were still operating [10]. Tar distillation took place on coal and, or coke works sites, and was the primary source of organic chemicals for different industries until petrochemical products took over in the 1960s [10]. The contamination at former gas and coke works varies with the range of products and by-products manufactured. On such sites, ground contamination arises from by-products, waste products from landfills and lagoons, and ancillary products such as ammoniacal liquor, coal tar, spent oxide and foul lime [12]. The organic contaminants are derived from constituents of coal tar such as aromatic hydrocarbons, polycyclic aromatic hydrocarbons (PAHs), phenils and phenols, nitrogen compounds, and organo – sulphur compounds, natural gas processing compounds such as alcohol, glycols, resins, heavy oils, and organic fuels such as petroleum and naphthalene [12]. Several of these coking works were developed on sites with a long industrial history, adding further contamination to that derived from the original activity. This study sets out to construct a conceptual model for the Avenue site that distinguishes local point source PAH16 (Polycyclic Aromatic Hydrocarbon) pollution arising from the coking plant activities, from the general historical diffuse pollution caused by other older industrial operations on the same site. Legacies of later phases of pollution that contribute to the same type of contamination can hide the special pattern of diffusion of contamination due to the earlier phase of industrial activity. One common problem in environmental risk assessment is to determine the value of a continuous attribute at any particular unsampled location; the uncertainty of any unsampled value; and the probability that a regulatory threshold for soil pollution or a criterion for soil quality is exceeded at any unsampled location, when only few sampled values are known [6, 7, 15, 16, 17, 19]. Geostatistics provide the basis for analysing data that vary continuously spatially and permit the inference values of the same variable at unsampled locations through interpolation techniques. Two key assumptions in geostatistical analysis are that (1) sample values are expected to vary continuously from one location to another; (2) at any particular location the value of the variable comprises a fixed component of the variation trend, which is usually unknown, and a random variable following one specific distribution [5, 15, 19] expressed by:

$$z(x) = a \oplus Z(x) \tag{1}$$

- $z(x)$ = value of the variable z at location x;
- a = fixed unknown component of the variation trend;
- $Z(x)$ = random variable described by:

$$Z(x) = m(x) + \varepsilon'(x) + \varepsilon'' \tag{2}$$

- $m(x)$ = deterministic function that describes the structural component with a constant mean or trend [3];
- $\varepsilon'(x)$ = a random but spatially correlated component, known as the variation of the regionalized variable, and it is the locally varying but spatially dependent residual of $m(x)$;
- ε'' = is residual, spatially independent noise, having a mean of zero and a standard deviation or variance σ^2.

If the assumption that an element varies continuously over a certain area is true, then it is customary to assume that the value at any point will be influenced much more by a closer known value of that element than by one farther away. Interpolation techniques are based on a normal or Gaussian distribution

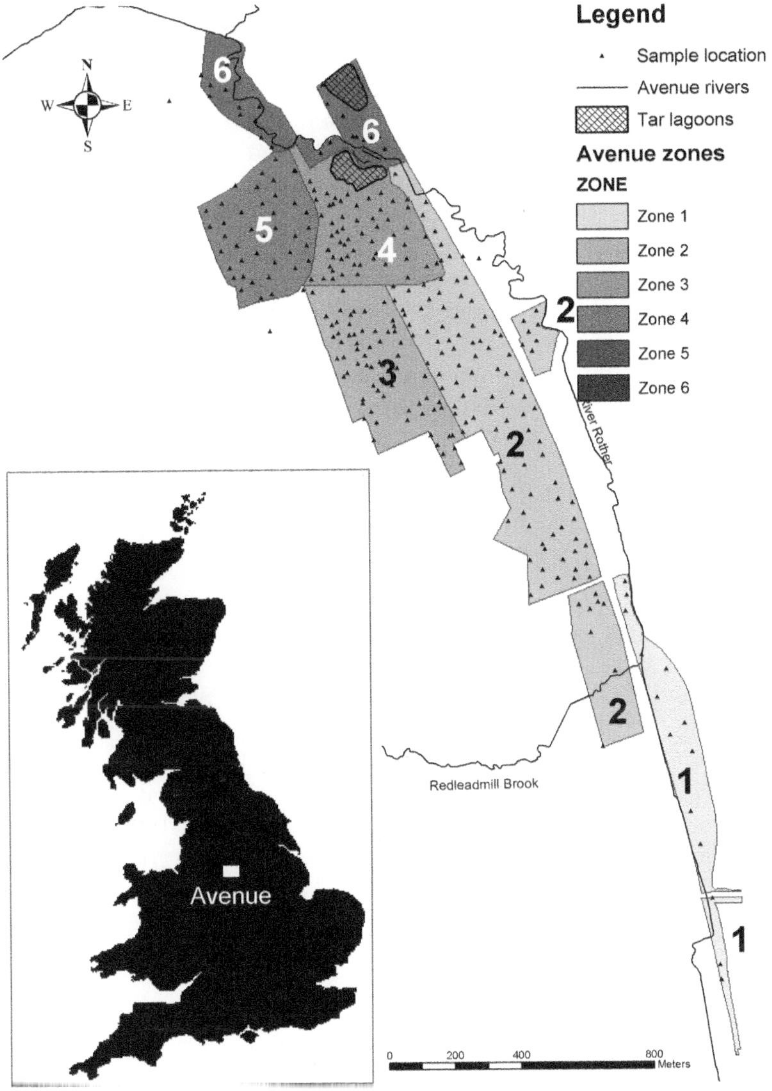

Fig. 1. Avenue zoning and soil sampling location

of data, and good spatial correlation. Problems arise when potentially continuous processes have not yet led to a normal spatial distribution, yet overlie an older continuous, but random process, which is spatially correlated.

Lark [21] models complex soil properties by assuming that the soil contamination is formed by a continuous but random component combined with a quasi point process. The quasi point process characterizes contamination (or any other process) in a small area of finite extent, which is represented by only one (or very few) soil sample(s) and does not diffuse continuously towards its neighbours. The continuous random processes are representative for the native metal content of the soil parent material and diffuse sources of pollution, while the quasi point process is defined by localized point sources of pollution. This situation may describe the pollution of an industrial site. If we consider that contamination with the same pollutant can result from both diffuse and point sources, we can expect its measured values to show very little spatial correlation, if any. These values, which have the point process values embedded, are called outliers and are considered unusual in their spatial context. The outliers do not belong to the continuous, but random, distribution of the majority of data, and are not necessarily extreme low or high values [2, 23]. In the case of pollution, if the outliers are not statistical anomalies due to errors in measurement or recordings, they indicate different processes superimposed on the same area and affecting the same variable [2, 18, 24].

2 Site Description

The colliery built in the 1880s at Avenue (Fig. 1), and the adjacent later lime and iron works were dismantled by 1938 and the site reverted to agriculture use. When the new, up-to-date 98 ha Avenue coking plant, built in the early 1950's to supply the Sheffield steel industry, was working at full capacity, it carbonised 2,175 tons of coal a day, producing 1,400 tons of smokeless fuel, 65 tons of 77% sulphuric acid, 35 tons of ammonium sulphate, 70,000 litres of crude benzole, and 250 tons of tar. Operations ceased in 1992 and since 1999 environmental reclamation work has been going on under the supervision of the Babtie Group [1].

3 Statistics

As part of the reclamation work, the site owners' consultants drilled 108 boreholes (BH) and 266 trial pits (TP). Seven hundred and twenty nine soil samples from depths between 10 cm and 18 m below surface level were analysed for PAH16. The PAH16 levels in parts of the site are two orders of magnitude higher than the PAH16 environmental threshold of 1,000 ppm. Overall the concentrations span from 0.05 ppm to over 20,000 ppm [1]. The soil samples were divided into four categories according to the depth of sampling, 0–1, 1–2, 2–4 m, and more than 4 m. It was originally hypothesised that the 265 soil samples between 10 cm and 1 m below surface level would be spatially correlated. To test this hypothesis the empirical PAH16 semi-variogram (Fig. 2) was built using the Geostatistical Analyst tools and ArcGIS 8.3, considering that it is more likely to have similar measured values close to the estimated

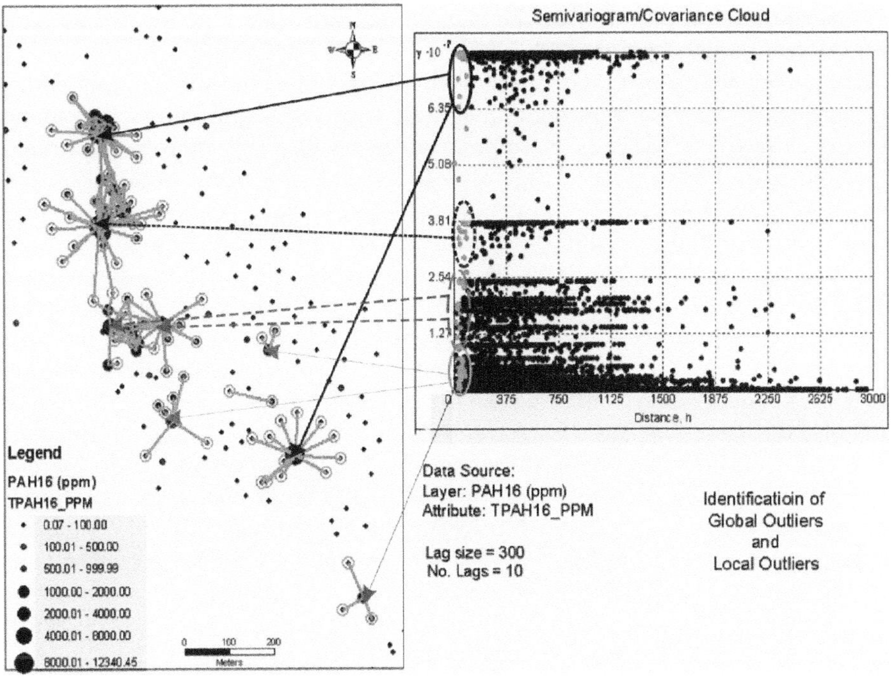

Fig. 2. Empirical semi-variogram for PAH16

point, but different measured values at further away. In this case the assumption is that the difference in values between two samples depends only on the distance between them and their relative orientation [5, 16]. In this case the variance, or standard deviation, of the sample value differences, varies only with the distance and the direction h between samples and it is known as a variogram.

$$2\gamma * (h) = \frac{1}{n}\sum_{1}^{n}[z(x) - z(x+h)]^2, \text{ where :} \tag{3}$$

- n = number of data pairs within a given class of distance and direction;
- $\gamma^*(h)$ = calculated semi-variance;
- $z(x)$ = value of the sample at location x;
- $z(x+h)$ = value of the sample at location $x+h$;

The results are plotted into a graph in which the horizontal axis represents the distance h and the vertical axis the experimental semi-variance, respectively. If two samples were picked from the same location, therefore h equals 0, we expect that the semi-variance value to be 0 for both calculated and measured semi-variance [5, 16]. The semi-variogram (Fig. 2) shows the presence of both global and local outliers for PAH16 and no spatial correlation. The global outliers are defined as very high or very low values relative with all the values in the dataset [2, 24], and in the semi-variogram they plot as distinct horizontal groupings of points [20]. The local outliers are values which, although not out of the dataset range, are abnormal relative to the surrounding values. Consequently, the local outliers have high semi-variogram values for pairs of points close to each other. These points plot close to the semi-variogram axis γ [2, 20, 23]. The arrows in Fig. 2 show the links between the local outliers in the empirical semi-variogram (pair of points very close in space with high semi-variance) and actual data locations in space (Avenue map).

A Q–Q normal plot for this data set suggests at least two different populations (Fig. 3a). This is interpreted as one population set modelling the diffuse pollution process, and the other describing the point source pollution process. The statistical outliers were identified through a box and whisker plot (Fig. 3b). This graph is depicting the first quartile, median, and the third quartile of a dataset. The box's lowest and highest horizontal limits represent the first and the third quartile positions on the y-axis, respectively. Fifty percent of the data values are plotted inside this box. The "whiskers" represent the percentile of the most extreme data-point, which is no more than 1.5 times the interquartile range from the box [25]. The outliers are values larger or equal to the sum of the third-quartile and the box interquartile range multiplied by 1.5 [26]. In Fig. 3b, the identified outliers are above the 80th percentile and range from 371 ppm to 12,340 ppm (49 samples out of 265, or 18.5% of the data) and represent the point source pollution from the coking works, while the remaining values represent the historical, diffuse contamination on the site. A Q–Q normal plot for these 49 outlier untransformed point source pollutant values suggests a single statistical population (Fig. 3c)

Spatially, the PAH16 outlier values cluster in three main areas, with a few isolated outliers elsewhere on the site (Fig. 4). The three main clusters are associated with (a) waste disposal and tar lagoon 4, or point source 1 (PS1), (b) stoking area, or point source 2 (PS2), and (c) main plant area, or point source 3 (PS3). The isolated outliers may be the result of individual spills or leakage from underground tanks.

The Avenue site was divided into Thiessen polygons based on the PAH16 sampling points (Fig. 5). This procedure took into account each sample depth,

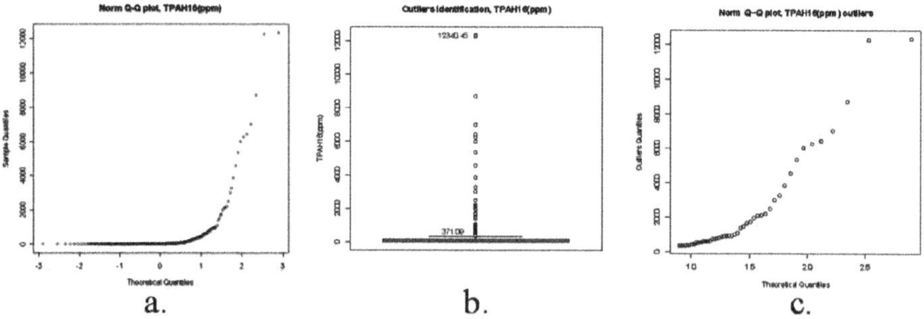

Fig. 3. **a**: normal Q–Q plot; **b**: box-whisker plot; **c**: normal Q–Q plot outliers

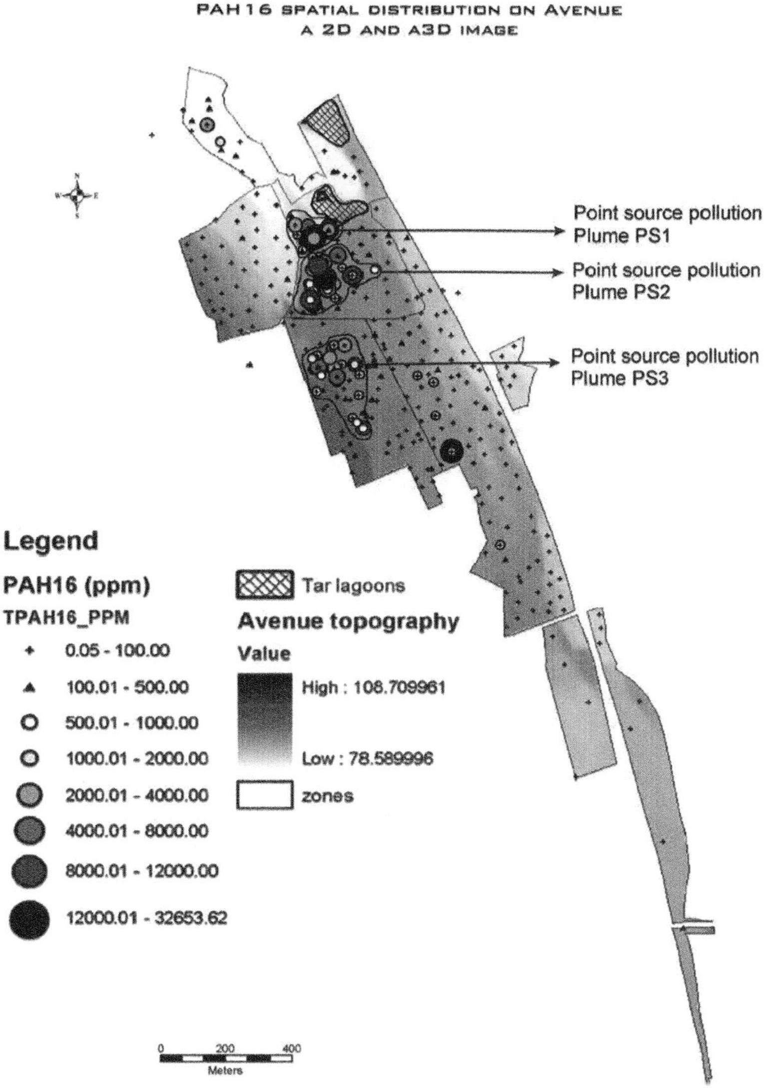

Fig. 4. Spatial distribution of PAH16

and it was assumed that each polygon has the characteristic of the sampled data. A narrow uncontaminated area of 12–18 m depth separates PS1 and PS2. Both PS2 and PS3 have small, unpolluted areas surrounded by high-polluted areas. Pollution may be present below those uncontaminated areas that were sampled only to a maximum depth of 2 m, since contaminated areas around them have very high PAH16 values below this depth. This may imply

Fig. 5. Identification of point source pollution

that uncontaminated material has been deposited on top of the contaminated ground.

A 3D model of source point pollution plume (PS1) based on 6 BH and 3 TP data, totalling 35 soil samples, was generated using a Thiessen polygon procedure (Fig. 6). Each sample represents the features of a 3D object with its Thiessen polygon as the base. The height of the object is obtained by taking the mid points between the sample depth and the sample depths above and

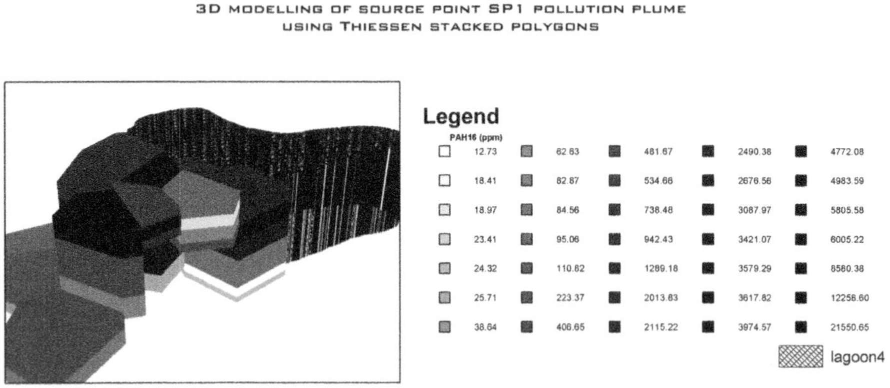

Fig. 6. 3D modeling of source point pollution plume PS1

below it. For the first and last sample depth, 20 cm were subtracted and added respectively, in order to also place these samples inside 3D objects. The PAH16 concentration decreases from tens of thousands ppm to a few thousands ppm to tens of ppm over just 2.5–5 m vertical distance. This demonstrates that not only the pollution is extremely localized, but also that diffusion and dilution of the point source pollution over 50 years has been relatively slight.

4 Conclusions

Superimposed processes over the same area contributing to variation in the same phenomenon can mask the spatial correlation of continuous variables. Identification of outliers helps to decide which data reflects the process that failed to diffuse over the entire study area and thus obliterate the initial spatial variability, but was significant enough to disturb the studied phenomenon's spatial correlation. The outlier dataset clusters spatially with the main clusters being associated with the waste disposal tip and tar lagoon 4, stocking area, and the main plant area. The remaining outliers probably represent the outcome of individual spills or leaks from underground tanks. The three-dimensional modelling of the pollution plume PS1 proves that the pollution diffusion and dilution over 50 years has been relatively insignificant. Three-dimensional modelling of pollution plumes combined with qualitative and historical data most likely will provide valuable information for remediation programmes, and increase our understanding of pollution processes.

Acknowledgements

I would like to thank all who have helped me in this work, particularly Babtie and East Midlands Development Agency, UK, for access to the site; Babtie, UK, for permission to use the site data; Nigel Lawson for help in gaining access to the site; and my supervisors: Prof. Ian Douglas and Prof. Robert Barr.

References

1. Babtie Group (2000) The Avenue Coking Works, Ground model & preliminary contamination assessment. Babtie Group, Fairbairn House, Ashton Lane, Sale, Manchester
2. Barnett V, Lewis T (1994) Outliers in Statistical data. Wiley series in probability and mathematical statistics, 3rd edn. John Wiley & Sons: New York
3. Burrough PA, McDonnell RA (1998) Principles of geographical information systems. Oxford University Press: Oxford
4. Carlon C, Critto A, Marcomini A, Nathanail P (2001) Risk based characterization of contaminated industrial site using multivariate and geostatistical tools. Environmental Pollution 111, pp 417–427
5. Clark I, (1978) Practical geostatistics. Elsevier Applied Science: New York
6. Clark and Harper (2000) Practical geostatistics 2000. Geostokos (Ecosse) Limited: Scotland, UK
7. Clark I, Harper WV (2001) Practical geostatistics 2000. Ecosse North America Llc.: Columbus, OH
8. D.E.T.R. (2000a) Contaminated land: Implementation of Part II A of the Environmental Protection Act 1990. London: HMSO
9. D.E.T.R. (2000b) Guidelines for environmental risk assessment and management. Revised Departmental Guidance. London: HMSO
10. D.o.E. (1987) Problems arising from the redevelopment of gas works and similar sites. Environmental Resources Limited, 2nd edn. London: HMSO
11. D.o.E. (1995a) Gas works, coke works and other coal carbonisation plants, Industry Profile. London: HMSO
12. D.o.E. (1995b) A guide to risk Assessment and risk management for environmental protection, London: HMSO
13. DTLR (2002) Development on land affected by contamination. Consultation paper on Draft planning technical advice, London: HMSO
14. Environment Agency (2002) Dealing with contaminated land in England, Progress in 2002 with implementing the part IIA regime, Environment Agency, Rio House, Bristol, UK
15. Goovaerts P (1997) Geostatistics for natural resources evaluation, Oxford University Press

16. Goovaerts P (1999) Geostatistics in soil science: state-of-the-art and perspectives, Geoderma 89, pp 1–45
17. Goovaerts P (2001) Geostatistical modelling of uncertainty in soil science, Geoderma 103, pp 3–26
18. Hawkins DM (1980) Identification of outliers, Monographs on Applied Probability and Statistics. Chapman and Hall
19. Issaks EH, Srivastava RM (1989) An introduction to applied geostatistics. Oxford University Press: New York, Oxford
20. Johnston K, Ver Hoef JM, Krivoruchko K, Lucas N (2001) Using ArcGIS geostatistical analyst, ESRI
21. Lark RM (2002) Modelling complex soil properties as contaminated regionalized variables, Geoderma 106, pp 173–190
22. Petts J, Cairney T, Smith M (1997) Risk-based contaminated land investigation and assessment. John Wiley & Sons Ltd.
23. Rousseeuw PJ, Leroy AM (1987) Robust regression and outlier detection. John Wiley & Sons, Ltd.
24. Shekhar SLC Zhang P (2003) A Unified approach to detecting spatial outliers, GeoInformatica 7:2, 139–166
25. The R Development Core Team (2003) The R environment for statistical computing and graphics, Reference Index, Version 1.8.0, http://www.r-project.org/, accessed 8/10/03
26. Tukey JW (1977) Exploratory data analysis. Addison-Wesley Publishing Company

A Note on Numerical Solutions of Least Squares Adjustment in GNU Project Gama

Aleš Čepek and Jan Pytel

Faculty of Civil Engineering, CTU Prague, Prague, Czech Republic
cepek@fsv.cvut.cz
pytel@fsv.cvut.cz

1 Introduction

Project Gama for adjustment of geodetic networks was started at the department of mapping and cartography, Faculty of Civil Engineering, Czech TU Prague, in 1998. Formerly it was planned to be only a local project with main goal to demonstrate students the power of object programming and at the same time to be a free independent tool for comparison of adjustment results from other sources. The Gama project received the official status of GNU software in 2001 and now contains a C++ library (including small C++ matrix/vector template library `gmatvec`) and two programs `gama-local` and `gama-g3`, that correspond to two development branches of the project.

Stable branch of the Gama project is represented by command line program `gama-local` for adjustment of three-dimensional geodetic networks in a local coordinates system. New development branch of the project (`gama-g3`) is aimed to the adjustment of geodetic networks in global geocentric system. The stable branch (`gama-local`) enables common adjustment of possibly correlated horizontal directions and distances, horizontal angles, slope distances and zenith angles, height differences, observed coordinates (used in sequential adjustment, etc.) and observed coordinate differences (vectors). Although such an adjustment model is now obsoleted by global positioning systems, it can still serve as an educational tool for demonstrating adjustment procedures to students and as a starting platform for the development of new branch of the project (`gama-g3`).

Numerical solution of least squares adjustment in geodesy is most commonly based on the solution of normal equations. As the Gama project was also meant to be a comparison tool, it was desirable to use a different method and Singular Value Decomposition (SVD) was implemented as the main numerical algorithm. As an testing alternative Gama implements another algorithm from the family of orthogonal decompositions based on Gram–Schmidt orthogonalization (GSO). Practical experience with both algorithms are discussed. In the Gama project geodetic input data are described in Extensible Markup Language (XML). The primary motivation for usage of XML was to define structured input data for adjustment of local geodetic network. The most important feature of XML is probably the ease of defining a grammar for user data (a class of XML documents) that consequently can be validated even independently of our applications. One of the goals of the Gama project is to build a collection of model geodetic networks described in XML. The lack of reliable testing data was one of major obstacles when testing implementation of numerical solution of the geodetic network adjustment.

2 Adjustment and Analysis of Observations

Geodesy as the scientific discipline is studying geometry of the Earth or, from the practical point of view, positioning if objects located on the Earth surface or in its relatively close boundaries. The input information is represented by geodetic *observations*.

The spectrum of observation types dealt by geodesy is very wide and ranges from classical astro-geodetic observations (astronomical longitude and latitude, variations and position of the Earth pole), measurements of geophysical quantities (gravity acceleration and its local anomalies), through traditional geometric observables like directions, angles and distances to photogrammetric measurements of historical monuments. But of the main importance in geodesy today are satellite global positioning systems (first of all NAVSTAR GPS and complementary other systems like DORIS or GLONASS).

The key role in processing of geodetic data belongs to the sphere of applied statistics in geodesy traditionally called *adjustment of observations*. The processing of geodetic observations is determined by the choice of appropriate mathematical model, which can be symbolically expressed as

$$\mathbf{f}(\mathbf{c}, \mathbf{x}, \mathbf{l}) = \mathbf{0}, \tag{1}$$

where $\mathbf{f}$ is a vector of functions describing relations between constants $\mathbf{c}$, unknown parameters $\mathbf{x}$ and observed quantities $\mathbf{l}$. Corresponding to the three components of this model are three mathematical spaces: parameter, observation and model space [1].

Three basic components of the mathematical model (1) are depicted in Fig. 1, where $\mathbf{A}, \mathbf{B}, \mathbf{G}$ and $\mathbf{H}$ are matrices of corresponding linearized relations (values of constants c are not estimated in geodesy and we can consider them to be a part of model space). Models can be direct, indirect or implicit; linear or nonlinear; can occur individually or in combinations

$$\begin{array}{lll} \text{model explicit in } \mathbf{x}: & \mathbf{x} = \mathbf{g}(\mathbf{l}), & \mathbf{x} = \mathbf{G}\mathbf{l} + \mathbf{v} \\ \text{model explicit in } \mathbf{l}: & \mathbf{l} = \mathbf{h}(\mathbf{x}), & \mathbf{l} = \mathbf{H}\mathbf{x} + \mathbf{v} \\ \text{implicit model}: & \mathbf{f}(\mathbf{x}, \mathbf{l}) = \mathbf{0}, & \mathbf{A}\mathbf{x} + \mathbf{B}\mathbf{l} + \mathbf{v} = \mathbf{0} \end{array}$$

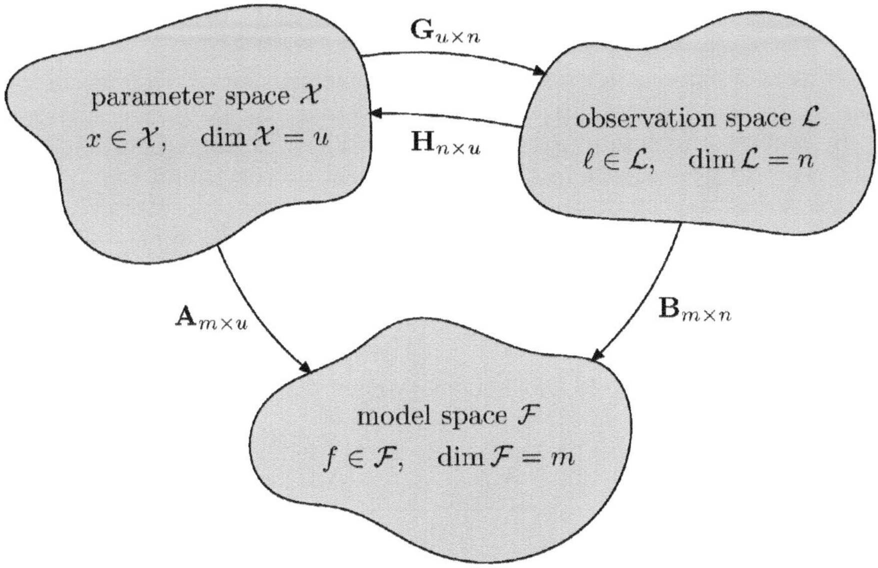

Fig. 1. Linear relations between parameter, observation and model spaces

3 Least Squares and Singular Systems

On adjustment of geodetic observations we are relatively often faced with models leading to singular sets of linear equations. Typically these are models without fixed points, ie. no points with fixed coordinates are given, or the number of fixed points is not sufficient (*free networks*, see [2] for more information).

Lets take as an example local network with observed directions and distances from Fig. 2. Relationship between unknown adjusted coordinates and observations can be expressed after linearization as the *project equations*

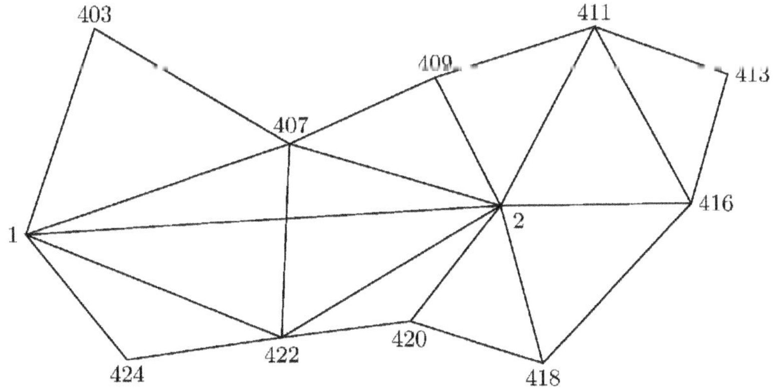

Fig. 2. Example of local geodetic free network

$$\mathbf{Ax} - \mathbf{l} = \mathbf{v} \tag{2}$$

where $\mathbf{A}$ is design matrix, $\mathbf{x}$ vector of unknowns, $\mathbf{l}$ vector of reduced observations and $\mathbf{v}$ vector of residuals (misclosure vector).

In geodesy the number of observation is always higher then number of unknowns. Project equations (2) thus represent overdetermined system and matrix $\mathbf{A}$ has more rows then columns. Least Squares is the basic method used in geodesy for observation adjustment, it gives us the unique solution $\mathbf{x}$ of system (2) that minimizes Euclidean norm of residual vector

$$\min \sqrt{\mathbf{v}'\mathbf{v}}. \tag{3}$$

A method commonly used for solving projects equations (2) (model explicit in observations) is based on *normal equations*

$$\begin{aligned} \mathbf{N} &= \mathbf{A}'\mathbf{A}, \qquad \mathbf{n} = \mathbf{A}'\mathbf{l}, \\ \mathbf{x} &= \mathbf{N}^{-1}\mathbf{n} \end{aligned} \tag{4}$$

Apart from unknown vector $\mathbf{x}$ (and residuals $\mathbf{v}$) we are always interested in geodesy in estimates of precision of adjusted quantities, in geodetic practice represented by variance-covariance matrix of adjusted unknowns $\mathbf{C}_{xx}$ and adjusted observations $\mathbf{C}_{ll}$

$$\mathbf{C}_{xx} = m_0'\mathbf{N}^{-1} \tag{5}$$

$$\mathbf{C}_{ll} = \mathbf{A}\mathbf{C}_{xx}\mathbf{A}' \tag{6}$$

The geometric shape of our adjusted network is defined by observed directions (or angles) and directions. If we fixed coordinates of two or more points the network shape would be necessarily distorted. Normal equations would lead to an adjustment solution in which residuals would be dependent on the coordinates of fixed points. This way we would degrade our observations in cases when coordinates of network points are either unknown or known with lower precision.

On the other hand if we consider coordinates of all points to be free, the corresponding matrix $\mathbf{N}$ is inevitably singular; columns of matrix $\mathbf{A}$ are linearly dependent (network can float freely in the coordinate system) and normal equation matrix $\mathbf{N}$ is positive-semidefinite

$$\mathbf{p}'\mathbf{N}\mathbf{p} \geq 0, \qquad \mathbf{p} \neq \mathbf{0}.$$

To get a unique solution we have to define additional constraints regularizing the system, preferably without deformation of the network shape. In geodetic practice we most often meet the following approaches

- Singular system is regularized by introducing pseudo-observations, typically with huge weights, that play a similar role as a set of constraint equations.

- Explicit system of constraint equations is defined to make the given system regular
$$\mathbf{Cx} = \mathbf{c}. \tag{7}$$
Normal equations then become
$$\begin{pmatrix} \mathbf{N} & \mathbf{C}' \\ \mathbf{C} & \mathbf{0} \end{pmatrix} \begin{pmatrix} \mathbf{x} \\ \mathbf{\Lambda} \end{pmatrix} = \begin{pmatrix} \mathbf{1} \\ \mathbf{c} \end{pmatrix}, \tag{8}$$
where $\mathbf{\Lambda}$ is the vector of Lagrange multipliers. In this case the matrix $\mathbf{C}$ is problem dependent and need to be known explicitly in advance.
- Euclidean norm of certain subset of unknown parameters vector $\mathbf{x}$ is minimized
$$\min_{x_i} \sqrt{\sum x_i^2}, \qquad i \in \mathcal{O}. \tag{9}$$
The set of indices $\mathcal{O}$ can contain all elements, but more often only selected elements of $\mathbf{x}$.

In the case of plane geodetic free network we can geometrically interpret the last constraint (9) as follows. By minimization of the Euclidean norm of residual vector (3) the shape and scale (if at least one distance is available) of the adjusted network together with covariances of adjusted observations are uniquely defined. The second additional constraint (9) then defines localization of the network in the coordinate system. Apart from the adjusted network shape we define simultaneously its shift and rotation in the coordinate system.

Another equivalent interpretation is that constraint (9) defines the particular solution of (2) in which the trace of variance-covariance submatrix corresponding to indices $i \in \mathcal{O}$ is minimal.

4 Normal Equations and Numerical Stability

Numerical solution of adjustment of observed quantities based on normal equations can be numerical unstable and in certain case we should prefer other numerical algorithms that solve directly the project equations (2). The possible source of troubles are the normal equations itself, or more precisely the condition number of normal equations. Let us restrict our discussion here to the simple case when matrix $\mathbf{A}$ does not contain linearly dependent columns and matrix $\mathbf{N}$ is positive-definite.

Condition number of matrix $\mathbf{A}$ is defined as
$$\kappa(\mathbf{A}) = \sqrt{\frac{\lambda(\mathbf{A}'\mathbf{A})_{\max}}{\lambda(\mathbf{A}'\mathbf{A})_{\min}}} \tag{10}$$
where $\lambda(\mathbf{A}'\mathbf{A})_*$ denotes maximal and minimal eigenvalue of matrix $\mathbf{A}'\mathbf{A}$. If we solve a linear set of equations then its condition number represents the

minimal upper estimate of the ratio of relative error of $\mathbf{x}$ and relative error of right hand side $\mathbf{l}$.

From the (10) directly follows that the condition number of normal equation matrix $\mathbf{N}$ is the square of condition number of the project equation matrix $\mathbf{A}$

$$\kappa(\mathbf{N}) = (\kappa(\mathbf{A}))^2 \tag{11}$$

We can say that when solving poorly conditioned normal equations we are loosing twice as much of correct decimal digits in a solution $\mathbf{x}$ compared with any direct solution of project equations.

Probably the most important class of algorithms for direct solution of project equations (2) is the family of orthogonal decomposition algorithms. Apart from other goals, GNU project Gama has been planned to be a kind of *etalon,* i.e. a tool for checking adjustment results from other software products. For this reason it was desirable to have adjustment based on a different numerical method other then traditional solution of normal equations and Singular Value Decomposition (SVD) was implemented as the main numerical algorithm. As an alternative another orthogonal decomposition adjustment algorithm GSO (based on Gram–Schmidt orthogonalization) is also available. We describe briefly both algorithm in the following section.

5 Gram–Schmidt Orthogonalization

Gram–Schmidt orthogonal decomposition is algorithm for computing factorization

$$\mathbf{A} = \mathbf{QR}, \qquad \mathbf{Q}'\mathbf{Q} = \mathbf{1} \tag{12}$$

where $\mathbf{Q}$ is orthogonal matrix and $\mathbf{R}$ is upper triangular matrix. Matrix $\mathbf{R}$ here is identical to the upper triangular matrix of Cholesky decomposition of normal equations

$$\mathbf{N} = \mathbf{A}'\mathbf{A} = \mathbf{R}'\mathbf{Q}'\mathbf{QR} = \mathbf{R}'\mathbf{R}. \tag{13}$$

Gram–Schmidt orthogonalization is a very straightforward and relatively simple algorithm that can be implemented in several variants differing in the order in which vectors are orthogonalized. The following three algorithms are adopted from [3, 300–301].

Algorithm 1.1 [Modified Gram–Schmidt (MGS) row version]

$$\begin{aligned}
&\text{for } k = 1, 2, \ldots, n \\
&\quad \hat{q}_k := a_k^{(k)}; \quad r_{kk} := (\hat{q}_k^T \hat{q}_k)^{1/2}; \\
&\quad q_k := \hat{q}_k / r_{kk}; \\
&\quad \text{for } i = k+1, \ldots, n \\
&\quad\quad r_{ki} := q_k^T a_i^{(k)}; \quad a_i^{(k+1)} := a_i^{(k)} - r_{ki} q_k; \\
&\quad \text{end} \\
&\text{end}
\end{aligned}$$

Algorithm 1.2 [Modified Gram–Schmidt (MGS) column version]

for $k = 1, 2, \ldots, n$
for $i = 1, \ldots, k-1$
$r_{ik} := q_i^T a_k^{(i)}; \quad a_k^{(i+1)} := a_k^{(i)} - r_{ik} q_i;$
end
$\hat{q}_k := a_k^{(k)}; \quad r_{kk} := (\hat{q}_k^T \hat{q}_k)^{1/2};$
$q_k := \hat{q}_k / r_{kk};$
end

Algorithms 1.3 [Classical Gram–Schmidt (CGS)]

for $k = 1, 2, \ldots, n$
for $i = 1, \ldots, k-1$
$r_{ik} := q_i^T a_k;$
end
$$\hat{q}_k := a_k - \sum_{i=1}^{k-1} r_{ik} q_i;$$
$r_{kk} := (\hat{q}_k^T \hat{q}_k)^{1/2}; \quad q_k := \hat{q}_k / r_{kk};$
end

It must not be forgotten that the variant known as *Classical Gram–Schmidt* has very poor numerical properties in that there is typically a severe loss of orthogonality among the computed q_i. Rearrangement of the calculation, known as *Modified Gram–Schmidt,* yields a much sounder computational procedure [4, pp. 230–232].

5.1 Generalized Orthogonalization Algorithm (GSO)

Generalized orthogonalization algorithm (GSO), a method based on Gram–Schmidt orthogonalization, for numerical solution of various adjustment models in geodesy was elaborated by František Charamza [5, 6]. GSO was implemented in GNU Gama to conserve this rarely used but interesting method and to represent an alternative numerical algorithm to SVD (which we expected to give better numerical results for numerically unstable systems).

Algorithm GSO operates on a block matrix structure

$$\begin{pmatrix} \mathbf{M}_1 & \mathbf{M}_2 \\ \mathbf{M}_3 & \mathbf{M}_4 \end{pmatrix} \longrightarrow \begin{pmatrix} \mathbf{Q}_1 & \mathbf{Q}_2 \\ \mathbf{Q}_3 & \mathbf{Q}_4 \end{pmatrix} \tag{14}$$

where transition from $\mathbf{M}$ to $\mathbf{Q}$ is defined by equations

$$\mathbf{Q'}_1\mathbf{Q}_1 = \mathbf{1} \tag{15}$$

$$\mathbf{M}_1 = \mathbf{Q}_1\mathbf{R} \tag{16}$$

$$\mathbf{Q}_1 = \mathbf{M}_1\mathbf{R}^{-1}, \quad \mathbf{Q}_2 = \mathbf{M}_2 - \mathbf{Q}_1\mathbf{Q'}_1\mathbf{M}_2 \tag{17}$$

$$\mathbf{Q}_3 = \mathbf{M}_3\mathbf{R}^{-1}, \quad \mathbf{Q}_4 = \mathbf{M}_4 - \mathbf{Q}_3\mathbf{Q'}_1\mathbf{M}_2 \tag{18}$$

and $\mathbf{R}$ is upper triangular matrix.

Algorithm MGS is applied to block matrix $\mathbf{M}$ so that column dot products are computed only for submatrices $(\mathbf{M}_1, \mathbf{M}_2)$, projections $r_{ki}q_k$ are computed for full columns of $\mathbf{M}$ and the whole process is terminated after processing of all columns of submatrix $(\mathbf{M}_1, \mathbf{M}_3)'$. This step is called *first orthogonalization* in algorithm GSO.

Lets take as an example system of project equations (2)

$$\mathbf{Ax} - \mathbf{l} = \mathbf{v}.$$

and apply algorithm GSO to the block matrix

$$\begin{pmatrix} \mathbf{A} & -\mathbf{l} \\ \mathbf{1} & \mathbf{0} \end{pmatrix} \longrightarrow \begin{pmatrix} \mathbf{Q} & \mathbf{v} \\ \mathbf{R}^{-1} & \mathbf{x} \end{pmatrix}$$

The result is directly the vector of unknown parameters $\mathbf{x}$ and vector of residuals $\mathbf{v}$. Cofactors (weight coefficients) of adjusted parameters $q_{x_i x_j}$ are available as dot products of rows i and j of submatrix $\mathbf{R}^{-1}$, cofactors of adjusted observations $q_{l_m l_n}$ are computed as dot products of rows m and n of submatrix $\mathbf{Q}$ and mixed cofactors $q_{x_i l_n}$ similarly as dot products of i-th row of $\mathbf{R}^{-1}$ and n-th row of matrix $\mathbf{Q}$.

5.2 Algorithm GSO and Singular Systems

Let us suppose now that project equations matrix $\mathbf{A}$ contains r linearly independent columns and remaining d linearly dependent columns. Without a loss generality we can assume that linearly dependent columns are located in the right part of matrix $\mathbf{A}$. We denote linearly independent columns $\mathbf{A}_1$, linearly dependent columns $\mathbf{A}_2$ and the matrix of their linearly combinations α

$$\mathbf{A} = (\mathbf{A}_1, \mathbf{A}_2), \qquad \mathbf{A}_2 = \mathbf{A}_1\alpha, \qquad \mathbf{x} = \begin{pmatrix} \mathbf{x}_1 \\ \mathbf{x}_2 \end{pmatrix} \tag{19}$$

Now we can rewrite project equations as

$$\mathbf{v} = \mathbf{A}_1\mathbf{x}_1 + \mathbf{A}_2\mathbf{x}_2 - \mathbf{l} = \mathbf{A}_1(\mathbf{x}_1 + \alpha\mathbf{x}_2) - \mathbf{l} = \mathbf{A}_1\tilde{\mathbf{x}} - \mathbf{l} \tag{20}$$

As the matrix $\mathbf{A}_1$ does not contain linearly dependent columns, the unique solution $\tilde{\mathbf{x}}$ of (20) exists that minimize Euclidean norm of $\mathbf{v}$.

If we know the matrix α and the vector $\tilde{\mathbf{x}}$ then any solution $\mathbf{x}$ of

$$\tilde{\mathbf{x}} = \mathbf{x}_1 + \alpha \mathbf{x}_2 = (\mathbf{1}, \alpha)\,\mathbf{x} \tag{21}$$

is at the same time the Least Square solution of (20) with the same vector of residuals $\mathbf{v}$.

If we apply algorithm GSO to the matrix

$$\mathbf{M}^{\mathrm{I}} = \left(\begin{array}{c|c} \mathbf{M}_1^{\mathrm{I}} & \mathbf{M}_2^{\mathrm{I}} \\ \hline \mathbf{M}_3^{\mathrm{I}} & \mathbf{M}_4^{\mathrm{I}} \end{array}\right) = \left(\begin{array}{c|cc} \mathbf{A}_1 & \mathbf{A}_2 & -\mathbf{l} \\ \hline \mathbf{1} & \mathbf{0} & \mathbf{0} \\ \mathbf{0} & \mathbf{1} & \mathbf{0} \end{array}\right) \tag{22}$$

we receive a block matrix

$$\mathbf{Q}^{\mathrm{I}} = \left(\begin{array}{c|cc} \mathbf{Q}_1 & \mathbf{0} & \mathbf{v} \\ \hline \mathbf{R}_1^{-1} & -\alpha & \tilde{\mathbf{x}} \\ \mathbf{0} & \mathbf{1} & \mathbf{0} \end{array}\right) \tag{23}$$

In the case of singular systems in GSO we have the *first orthogonalization,* that defines particular solution in which the unknowns parameters corresponding to linearly dependent columns of $\mathbf{A}$ are set to zero. From CGS directly comes out that matrix α is the matrix of linear combinations from (19). Cofactors are computed the same way as in the case of regular systems.

On numerical computation of GSO we naturally do not obtain exactly zero vectors on positions of (almost) linearly dependent columns. We declare to be linearly dependent those columns of $\mathbf{A}$ whose norms drop under a given tolerance. During the first orthogonalization we set to zero corresponding subvectors in the area of $\mathbf{A}_2$. These values can be considered just a *random noise* that adds no information to the whole solution.

The result of first orthogonalization are first of all the vector of residuals and cofactors of adjusted observations. Now remains to determine the vector of unknown parameters $\mathbf{x}$ that satisfies condition (9) and its cofactors (weight coefficients). This step of GSO is called *second orthogonalization.*

By solving the system of linear equations

$$\begin{pmatrix} -\alpha \\ \mathbf{1} \end{pmatrix} \mathbf{x}_2 = \begin{pmatrix} \tilde{\mathbf{x}} \\ \mathbf{0} \end{pmatrix} \tag{24}$$

we get, according to (21), a vector $\mathbf{x}$ with minimal norm. If we select from (24) only certain rows, we obtain the solution minimizing the corresponding subvector. This system can naturally be solved using GSO.

If we need cofactors of adjusted unknowns, as is the standard case with geodetic applications, we have to process during second orthogonalization the whole lower submatrix that resulted from the first orthogonalization step.

$$\mathbf{M}^{\mathrm{II}} = (\mathbf{M}_1^{\mathrm{II}} \mid \mathbf{M}_2^{\mathrm{II}}) = \left(\begin{array}{c|cc} -\alpha & \mathbf{R}_1^{-1} & \tilde{\mathbf{x}} \\ \mathbf{1} & \mathbf{0} & \mathbf{0} \end{array}\right) \tag{25}$$

During the first orthogonalization linearly dependent columns in $\mathbf{M}_1$ are identified and are explicitly zeroed. The results of first orthogonalization is a particular solution in which unknowns corresponding to linearly dependent columns are all set to zero. Naturally their cofactors are zero as well.

During the second orthogonalization step only the submatrix $(\mathbf{Q}_3^{\mathrm{I}}, \mathbf{Q}_4^{\mathrm{I}})$ is influenced and the orthogonalization process is carried as follows

- Gram–Schmidt orthogonalization runs only through columns corresponding to linearly dependent columns of $\mathbf{M}_1$ as if they were numbered $1, 2, \ldots, d$, where d is the nullity of $\mathbf{M}_1$
- dot products are computed only for indexes $i \in \mathcal{O}$ from the regularization condition

$$\min_{x_i} \sqrt{\sum x_i^2}, \qquad i \in \mathcal{O}$$

Linearly dependent columns are zeroed during second orthogonalization even in the region of submatrix $(\mathbf{Q}_3, \mathbf{Q}_4)$. Cofactors after second orthogonalization are computed the same way as in the case of regular systems.

6 Singular Value Decomposition (SVD)

For any real $m \times n$ matrix $\mathbf{A}$, $m \geq n$, there exists singular value decomposition

$$\mathbf{A} = \mathbf{UWV}' \tag{26}$$

$$\mathbf{U}'\mathbf{U} = \mathbf{1} \qquad \mathbf{VV}' = \mathbf{V}'\mathbf{V} = \mathbf{1}$$

where $\mathbf{U}$ is $m \times n$ matrix with orthogonal columns, $\mathbf{W}$ is diagonal matrix $n \times n$ with nonnegative elements and $\mathbf{V}$ is square orthogonal matrix $n \times n$. (this variant is referred to as the *thin SVD* [4]).

The matrix $\mathbf{W}$ is uniquely determined up to the permutation of its diagonal elements. Diagonal elements w_i are called singular values of matrix $\mathbf{A}$. Their squares are eigenvalues of $n \times n$ matrix $\mathbf{A}'\mathbf{A}$. Thus, the condition number of matrix $\mathbf{A}$ can be compute as ratio of maximal and minimal singular value.

$$\kappa(\mathbf{A}) = \frac{w_{\max}}{w_{\min}} \tag{27}$$

With singular decomposition we can directly express the vector of unknown parameters $\mathbf{x}$ from project equations

$$\mathbf{Ax} = \mathrm{l}, \qquad \mathbf{x} = \mathbf{VW}^{-1}\mathbf{U}'\mathrm{l}, \qquad \mathbf{W}^{-1} = \mathrm{diag}(1/w_i) \tag{28}$$

If matrix $\mathbf{A}$ has more rows then columns (overdetermined system), then the Euclidean norm of residual vector

$$\mathbf{v} = \mathbf{Ax} - \mathbf{l}$$

is minimal and the vector $\mathbf{x}$ is the Least Squares solution to project equations (2).

For a matrix $\mathbf{A}$ with linearly dependent columns d singular values are zero (d is dimension of null space of $\mathbf{A}$). Singular value decomposition explicitly constructs orthonormal vector basis of the null space and the range of $\mathbf{A}$. Columns of the matrix $\mathbf{U}$ corresponding to nonzero singular values w_i form the orthonormal base of the range of $\mathbf{A}$. Similarly columns of matrix $\mathbf{V}$ corresponding to nonzero singular values form the orthonormal basis of the null space of $\mathbf{A}$.

$$\begin{aligned}\mathcal{N}_{\mathbf{A}} &= \{\mathbf{x} \mid \mathbf{Ax} = \mathbf{0}, \quad \mathbf{x} \in \mathcal{R}^n\} \\ \mathcal{R}_{\mathbf{A}} &= \{\mathbf{y} \mid \mathbf{y} = \mathbf{Ax}, \quad \mathbf{x} \in \mathcal{R}^n\}\end{aligned}$$

In the case of rank deficient systems, we set into the diagonal of inverse matrix $\mathbf{W}^{-1}$ zeros instead of reciprocals for elements corresponding to linearly dependent columns $\mathbf{A}$

$$\mathbf{W}^{-1} = \operatorname{diag}\begin{cases} 1/w_i & \text{pro } w_i > 0 \\ 0 & \text{pro } w_i = 0 \end{cases} \tag{29}$$

Resulting particular solution $\mathbf{x}$ minimizes both Euclidean norm of residuals and at the same time the norm of unknown parameters $\mathbf{x}$.

Rather surprising replacement of reciprocal $1/0 \equiv \infty$ by zero can be explained as follows. Solution vector $\mathbf{x}$ of overdetermined system

$$\mathbf{Ax} = \mathbf{l}$$

can be expresses as the linear combination of columns of matrix $\mathbf{V}$

$$\mathbf{x} = \sum_{i=1}^{n} \left(\frac{1}{w_i}\mathbf{U}_{(i)}\mathbf{l}\right) \mathbf{V}_{(i)} \tag{30}$$

Coefficients in the parenthesis are dot products of columns $\mathbf{U}$ and right hand site $\mathbf{l}$ multiplied by reciprocal value of the singular value. Zero singular values correspond to linearly dependent columns of matrix $\mathbf{A}$ that add no other information to the given system. Setting corresponding diagonal elements of matrix $\mathbf{W}^{-1}$ to zeros is equivalent to elimination of linearly dependent columns from the matrix $\mathbf{A}$.

With matrix $\mathbf{W}^{-1}$ defined according to (29), cofactors are computed the same way for regular and singular systems

$$\mathbf{Q}_{xx} = \mathbf{N}^{-1} = (\mathbf{A}'\mathbf{A})^{-1} = (\mathbf{VW}'\mathbf{U}'\mathbf{UWV}')^{-1} = \mathbf{VW}^{-1}\mathbf{W}^{-\mathrm{T}}\mathbf{V}' \tag{31}$$

$$\mathbf{Q}_{ll} = \mathbf{AQ}_{xx}\mathbf{A}' = (\mathbf{UWV}')(\mathbf{VW}^{-1}\mathbf{W}^{-\mathrm{T}}\mathbf{V}')(\mathbf{VW}'\mathbf{U}') = \mathbf{UU}' \tag{32}$$

$$\mathbf{Q}_{lx} = \mathbf{AQ}_{xx} = \mathbf{UWV}'\mathbf{VW}^{-1}\mathbf{W}^{-\mathrm{T}}\mathbf{V}' = \mathbf{UW}^{-1}\mathbf{V}' \tag{33}$$

Cofactors (weight coefficients) for adjusted parameters, observations and mixed cofactors are computed, similarly is in the case of GSO, as the dot products of rows of matrices $\mathbf{U}$ and $\mathbf{V}$; multiplied by diagonal elements of $\mathbf{W}^{-1}$ in the case of cofactors of $\mathbf{x}$.

6.1 Algorithm SVD and Singular Systems

What now remains is to show how to compute the particular solution that minimizes only a given subset of subvector $\mathbf{x}$ according to the second regularization condition (9). We compose overdetermined system of linear equations

$$\psi\mathbf{c} + \mathbf{x} = \hat{\mathbf{x}} \tag{34}$$

where columns of matrix ψ are vectors of null space basis

$$\psi = \left(\mathbf{V}_{(i_1)}, \mathbf{V}_{(i_2)}, \ldots, \mathbf{V}_{(i_d)},\right), \qquad w_{i_n} = 0$$

and $\mathbf{c}$ is the vector of coefficients of linear combination of null space basis vectors that, when added to vector $\mathbf{x}$, minimizes selected subvector of unknown parameters $\hat{\mathbf{x}}$ (here they act as residuals).

From comparing (34) with (24) and (25) it is obvious that for computing $\hat{\mathbf{x}}$ we can use second orthogonalization of algorithm GSO. If the GSO second orthogonalization is applied to the matrix $\mathbf{V}$ from singular decomposition

$$\mathbf{M}^{\mathrm{II}} = \left(\mathbf{M}_1^{\mathrm{II}} | \mathbf{M}_2^{\mathrm{II}}\right) = \left(\psi | \boldsymbol{\Psi}\right), \tag{35}$$
$$\boldsymbol{\Psi} = \left(\mathbf{V}_{(j_1)}, \mathbf{V}_{(j_2)}, \ldots, \mathbf{V}_{(j_r)},\right), \qquad w_{j_k} \neq 0$$

we obtain a matrix $\hat{\mathbf{V}}$. If we now replace singular value decomposition matrix $\mathbf{V}$ by the matrix $\hat{\mathbf{V}}$, we can compute vector $\hat{\mathbf{x}}$ and all cofactors according to the same formulas (30), (31), (32) and (33) as in the case of standard SVD solution $\mathbf{x}$.

7 Network Adjustment in GNU Gama

Gama was started in 1998 as a local educational project, mainly to demonstrate our students the power and capability of object programming (the project is written in C++) and at the same time to show some alternatives to traditional approach of numerical solutions of Least Squares adjustments based on normal equations. Project Gama was released under the terms of GNU General Public license and in 2001 received the official status of GNU software.

Numerical solution of geodetic network adjustment in Gama is based on an abstract C++ class and currently two derived classes are available implementing algorithms SVD and GSO. SVD is the primary algorithm used in Gama (one of our long term goals is to add more numerical solutions, namely solutions exploiting sparse structure of project equations). From this perspective algorithm GSO was implemented in Gama only as an testing alternative, either for comparing numerical results and for verification of the adjustment classes hierarchy in practice.

It is generally agreed that a bad implementation of GSO can produce disastrous results. For example, during first orthogonalization step of GSO we set to zeros unknown parameters corresponding to linearly dependent columns. In the case of free geodetic network adjustment these are coordinates of some points—the whole network is *pinned* on these points and clearly, if close points are selected the regularization is unstable. The order of columns in orthogonalization is important.

From practical experience we know that vector norms in GSO orthogonalization process generally tend to decrease. As GSO is just an alternative algorithm in Gama and its performance is not a crucial point, we implemented it with full pivoting, ie in each orthogonalization cycle the vector with maximal norm is selected as a pivot (with this modification GSO is about twice as slow compared to SVD for large networks).

Singular Value Decomposition is a very robust method for dealing with systems that are either singular or numerically close to singular. Even with *full pivoting* we expected GSO to prove to be inferior compared to SVD, at least in cases with ill-conditioned matrices. Surprisingly, with all real geodetic networks we have available this was not the case. Apart from real data we used for testing series of random generated three-dimensional networks.

Our implementation of SVD is based on a classical algorithm published by Golub and Reinsch [7] (the ALGOL procedure SVD). The decomposition is constructed in two phases. It starts with Householder reduction to bidiagonal form followed by diagonalization. Contrary to our expectations, SVD as used in Gama has not proved to give numerically better results and in some cases it even lost convergence in the diagonalization phase.

Simple and tempting explanation, that comes first to mind, would be that SVD implementation in Gama is somehow wrong. After all testing and revisions this does not seem to be the point. A possible explanation might give us the following quotation from [4]

> ... Finally, we mention Jacobi's method ... for the SVD. This transformation method repeatedly multiplies A on the right by elementary orthogonal matrices (Jacobi rotations) until A converges to $U\Sigma$; the product of the Jacobi rotations is V. Jacobi is slower than any of the above transformation methods (it can be made to run within about twice the time of QR ...) but has the useful property that for certain A it can deliver the tiny singular values, and their singular vectors, much more accurately than any of the above methods provided that it is properly implemented ...

Surely to have more numerical methods implemented in Gama would be helpful, for example the above mentioned Jacobi's method for SVD.

A practical problem, during testing of the adjustment methods in Gama, was relative shortage of reliable observation data and their adjustment results for testing. To enable easy comparison with other softwares we defined description of geodetic networks in XML (we use DTD for the definition of the formal syntax of our structured data). Conversion from a well defined data format into XML is relatively simple task but processing of XML is not a trivial task and cannot be done without a XML parser. In GNU Gama project we use XML parser `expat` by James Clark, see `http://expat.sourceforge.net/`. We believe that XML is the best data format for description and exchange of structured data in Gama project. One of the goals of our project is to compile a free collection of geodetic networks described in XML.

References

1. Petr Vaníček and Edward J. Krakiwsky (1986) Geodesy: The Concepts, 2nd ed., North-Holland, Amsterdam
2. Karl-Rudolf Koch (1999) Parameter Estimation and Hypothesis Testing in Linear Models, 2nd ed., Springer-Verlag, Berlin
3. Åke Björck (1994) Numerics of Gram–Schmidt Orthogonalization, Linear Algebra and Its Applications 197, 198:297–316
4. Gene H. Golub and Charles F. Van Loan (1996) Matrix Computations, 3rd ed., The John Hopkins University Press, Baltimore
5. Charamza, F. (1979) GSO—An Algorithm for Solving Least-Squares Problems with Possibly Rank Deficient Matrices, Optimization of Design and Computation of Control Networks, Akadémiai Kiadó, Budapest
6. Charamza, F. (1978) An Algorithm for the Minimum-Length Least-Squares Solution of a Set of Observation Equations, Studia geoph. et geod., Vol 22, pp. 129–139
7. G. H. Golub and C. Reinsch (1971) Singular Value Decomposition and Least Squares Solutions, Numer. Math. 14, 403–420 (1970), Handbook for Auto. Comp., Vol II—Linear Algebra, 134–151.

Presentation of Entrepreneurship Data and Aspects of Spatial Modeling

Robert J. Breitenecker[1], Jürgen Pilz[2], and Erich J. Schwarz[3]

[1] Department of Innovation Management and Entrepreneurship, University of Klagenfurt, Klagenfurt, Austria
robert.breitenecker@uni-klu.ac.at

[2] Department of Statistics, University of Klagenfurt, Klagenfurt, Austria
juergen.pilz@uni-klu.ac.at

[3] Department of Innovation Management and Entrepreneurship University of Klagenfurt, Klagenfurt, Austria,
erich.schwarz@uni-klu.ac.at

1 Introduction

Positive effects of new firms on the job market, technology transfer, and contributions to structural change has turned political attention to start-ups. Every year the number of new firm creations increases, where on the other hand the number of major enterprises decreases.[4] More and more firms have no employees and a trend to small-scale self employment can be recognized.[5] Political and economic support programs try to revoke regional discrepancies of business activity, firm development and foundation activity. These programs have to be evaluated and improved continuously.

An meaningful curatorial foundation statistic for entrepreneurship research and a statistic to assess the entrepreneurial activity in Austria for political decision making does not exist until now.[6] Solely the Federal Economic Chamber of Austria (WKO) reports a statistic of foundation activity of commercial firms every year.[7] This statistic permits to observe a trend of firm foundations,

[4] Cp. Wirtschaftskammer Österreich [16]: 23.

[5] Cp. Schwarz and Grieshuber [11]: 103ff.

[6] For the statistical situation in Germany see e.g. Fritsch et al. [3]: 2f. For the effort to the development of the curatorial statistical system in Germany cp. Struck [14]: 41ff.

[7] The Federal Economic Chamber of Austria is the legal representation of interests of Austrian entrepreneurs. In its founding statistic the number of new start-ups are calculated from new entrants into the membership database of the WKO. To exclude pseudo foundations and multiple data set entries the database has been revised. A detailed description of data revision can be found in Wirtschaftskammer Österreich [16].

but does not map the overall Austrian foundation activity. Firms which are not in the scope of the WKO are not included in this statistic.[8] Few other Austrian public services cumulate data from newly founded firms, but the access to these data sources is limited and the data is not appropriate for research.[9]

To ensure a continuous and complete evaluation of supporting programs it would be wise to design a monitoring system for all Austrian enterprises which includes all commercial and noncommercial firm foundations and closings. Such a system would be a valuable source for entrepreneurship research, but suitable statistical measures and methods for presenting and modeling non-normal spatial entrepreneurship data are needed to build such an overall monitoring system.

This paper briefly reports measures and methods commonly used in descriptive statistics for presenting entrepreneurship data in reference to its spatial distribution. We illustrate these by an example of numbers of new start-ups in Austria in the year 2001.[10] By applying different measures we show how sensible presenting regional differences in foundation activity can be. Further we will give a brief introduction of spatial general linear models [5] and the hierarchical Bayesian models for count data [15] for modeling non-normal spatial data.

2 Presenting and Mapping Entrepreneurship Data

Charts and tables are instruments for descriptive and explorative data analysis. They are helpful in visualizing data, building hypotheses and presenting results from statistical computation with spatial reference. To compare regional discrepancies, the right measure has to be specified to include different area or population of regions in the calculation or presentation.

For example if the differences in firm foundation activity of Austrian provinces are to be compared, the absolute numbers or the percentage of counted foundation will not be practical. Regions with different size and population cannot be compared with non standardized measures. Although the absolute values and the percentage are improper measures, both can be found in regional comparisons.[11]

We give an example of how different measures of foundation activity can influence the ranking of regions. Figure 1 shows the number of firm foundations

[8] Foundations in the field of agriculture and forestry and freelancers which belong to another chamber or not, are not registered by this and any other official statistic.

[9] E.g. social insurance institution, finance office, commercial credit agencies.

[10] Data from Wirtschaftskammer Österreich [16].

[11] On the web site of the Lower Austria's Business Portal the foundation statistic of WKO can be found, but on the site only the absolute counts and the percentage to all new Austrian firms for 2002 are presented (www.loweraustria.biz/upload/downloads/Betriebsgruendungen%202002.pdf).

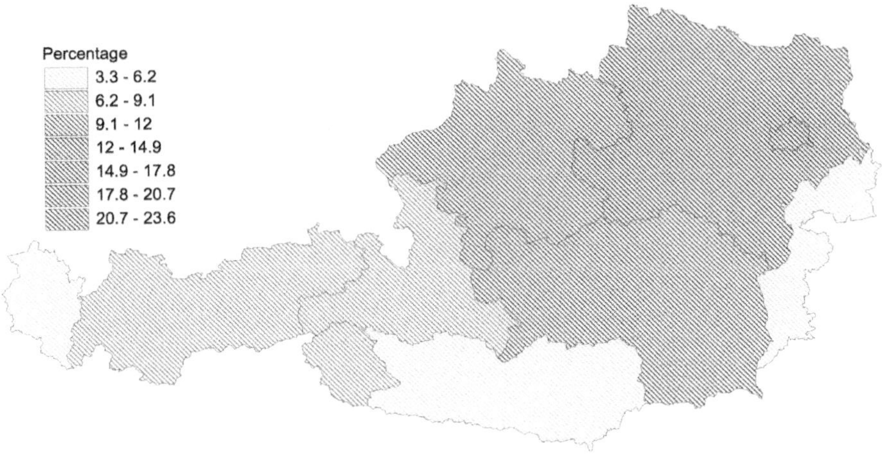

Fig. 1. Percentage of new firm foundations in Austrian provinces 2001

of Austrian provinces in percent to all new start-ups in Austria in a map. In the year 2001, Vienna is ranked first with 6,145 new firm foundations or a percentage of 23.1%, before Lower Austria with 5,351 new firms or 20.6%. It follow Styria, Upper Austria, Tyrol, Salzburg, Carinthia and Vorarlberg. The province of Burgenland with 869 or 3.3% of new enterprises of Austria is found at the lower end in this statistic.[12] We will see how the ranking of provinces (especially Upper Austria and Burgenland) concerning foundation activity will change when applying two different standardized measures.

A standardized measure for presenting the number of firm foundations on the level of provinces is the foundation intensity, presented in the foundation statistic of the WKO.[13] The intensity is the percentage of start-ups in relation to all active members of the WKO in the according region, where active members are all existing and active firms registered by the WKO. Figure 2 shows that Burgenland is now on the fourth place with a percentage of 9.2% new firms above the average of Austria with 8.7%. Upper Austria is with 7.4% below average only on the sixth place.[14]

Egeln et al. [2] calculate a measure for start-ups in relation to the potential of foundations, where the potential founders are the employees of the observed region and province, respectively. Start-ups per employees is the number of firm foundations in relation to 1,000 employees in the region.[15] Figure 3 shows this statistic. Compared to Figs. 1 and 2, differences in ranking of the nine

[12] Cp. Wirtschaftskammer Österreich [16]: 21.

[13] Cp. Wirtschaftskammer Österreich [16]: 22; Fritsch and Niese [4]: 4. Fritsch and Niese calculate the number of new start ups in relation to 100 existing firms of the respective region.

[14] Cp. Wirtschaftskammer Österreich [16]: 22.

[15] Cp. Egeln et al. [2]; Fritsch and Niese [4]: 3f.

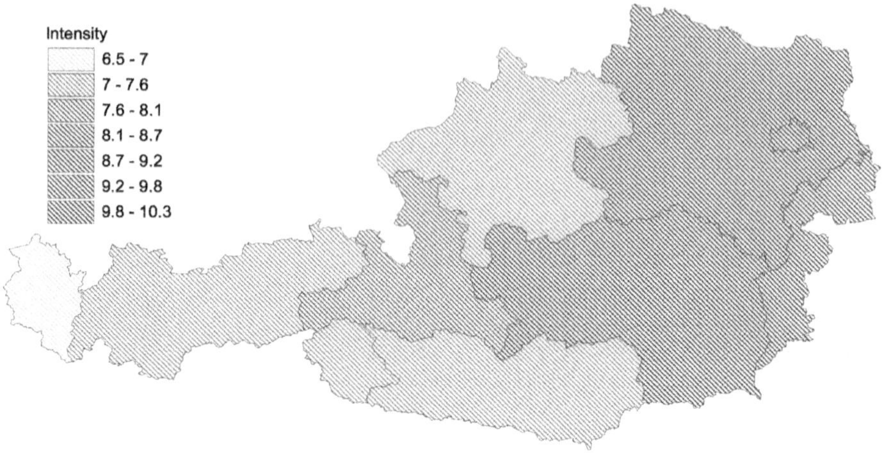

Fig. 2. Start-ups per 100 existing firms in Austria 2001

Austrian provinces can be recognized. Now Burgenland is ranked first, Upper Austria is ranked at last.[16]

But also this measure is not eligible for regional comparison, because every province has a different industrial history and therefore a different firm structure. Figure 4 shows the number of employees per working place in the nine provinces, from which the firm structure in the provinces can be deduced.

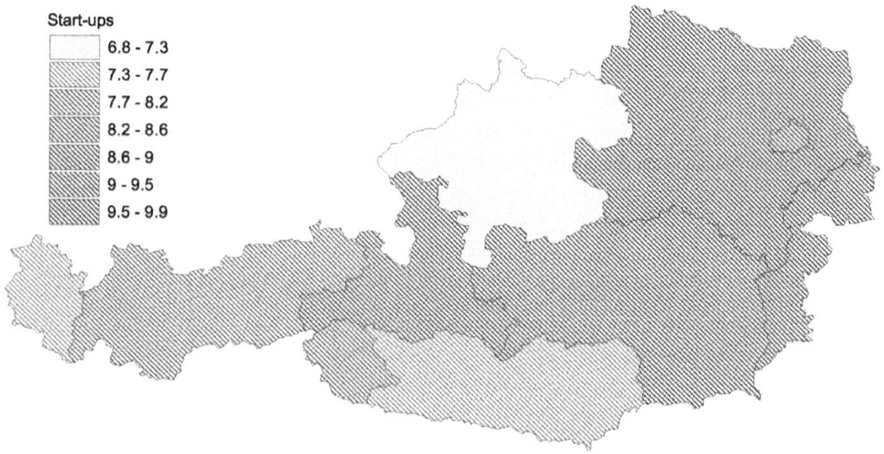

Fig. 3. Start-ups per 1,000 employees in Austria 2001

[16] Data from Wirtschaftskammer Österreich [16]: 21 and Statistik Austria [13] with own calculations.

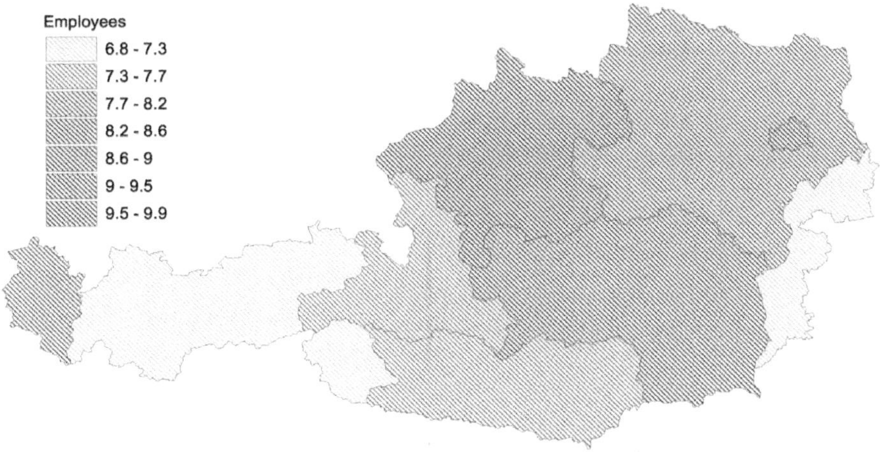

Fig. 4. Employees per place of work in Austria 2001

Upper Austria with its iron, steel and chemical industry has many major enterprises. Burgenland seams to have more small enterprises.[17]

This short example shows that choosing the right measure for such data is very sensitive, particularly if political decisions have to be made on the basis of such statistics. Area, population and firm structure should be considered by comparing different countries or provinces.

To make statistical information from different EU countries and regions comparable, EUROSTAT has established the Nomenclature des unites territoriales statistiques (NUTS). Using NUTS classification ensure that regions of comparable size all appear at the same level and making it possible to compare. Each NUTS unit contains regions which are similar in terms of area, population, economic weight or administrative power.[18] In Table 1 eigth different levels for presentation and comparison of regions in Austria, including the NUTS units, are presented.[19] It shows the configuration and the number of regions of these units.

With the exception of the post code, all lower levels can be aggregated to a higher level. To assign new firm foundation activity to the levels of NUTS 3, political districts or communities, the addresses or the post codes of the firms can be used. A main problem in that case is that classification with the post codes to these levels is not unique. The range of post districts overlap

[17] Data Source: Statistik Austria [13].

[18] For more information about NUTS see EUROSTAT on the web (http://europa.eu.int/comm/eurostat/ramon/nuts/splash_regions.html).

[19] Data from http://www.statistik.at/fachbereich_topograph/tab2.shtml for political units in Austria, http://www.statistik.at/verzeichnis/nuts.pdf for NUTS classification and http://www.statistik.at/verzeichnis/gemeindeverzeichnis.shtml to count post regions.

Table 1. Regional units for presentation in Austrian maps

unit	regions of unit	no. of regions
NUTS 0	Country – Austria	1
NUTS 1	Groups of Provinces	3
NUTS 2	Austrian Provinces	9
NUTS 3	Groups of Political Districts	35
district	Austrian Political Districts	99
post code	Post Regions	2,073
community	Austrian Communities	2,359
locality	Austrian Localities	17,364

with administrative boundaries. To ensure a unique classification, the total firm address, with the knowledge in which community it falls, has to be used.

An alternative to the regional levels is to assign the firm address to geographic coordinates (x, y). Different commercial providers offer the service to assign addresses to geographic coordinates.[20] A subsequent aggregation to other levels can be done ex post.

3 Spatial Model Building for Entrepreneurship Data

By presenting count data in epidemiology it is common practice to smooth over regions to avoid extreme values for regions with few observations or less population. Such extreme values are hard to interpret and lead to misinterpretations. With kernel smoother geographical trends and implications will be more trusty.[21] Considering the example of new start ups the foundation rate of region i will be calculated as weighted sum of the rates of neighbored regions.[22]

There are several possibilities to calculate spatial dependences and neighborhoods between coordinates and/or regions, but it should be decided whether these methods make sense applied to entrepreneurship data. A basic method to define the neighborhood between two coordinates is the euclidean distance. For calculating this or any other metric between regions, the geographic centers, central places or the main cities of these regions have to be defined. The method to define neighborhoods of regions over common borders is well known in epidemiology. For epidemiological and environmental data these methods seem to be acceptable but don't seem to be for entrepreneurship data. It makes no sense to define the distance between two regions or enterprises by a straight line, when practically streets and/or mountains have

[20] E.g. WIGeoGIS for Austria (http://www.wigeogis.com).

[21] Cp. Koboltschnig [8]: 9.

[22] A detailed introduction in kernel density estimation can be found in Silverman [12].

to be negotiated or two regions with common borders don't have to share one street. On this account an alternative defining neighborhoods between regions could be common infrastructure. In this case two regions are neighbors, if they have a street, a highway or a railroad line in common. For a detailed illustration of different neighborhood definitions and different spatial linear modeling strategies in the field of entrepreneurship research see Breitenecker [1].

3.1 Generalized Linear Model for Spatial Prediction

Gotway and Stroup [5] introduce a spatial approach for analyzing non-normal data. We give a brief introduction how in terms of Gotway and Stroup [5] the theory of generalized linear models can be extended to include discrete and categorical data for spatial prediction.

Let $\underline{Z} = (Z(s_1), \ldots, Z(s_n))'$ be a vector of random variables, each having a distribution in the exponential family, and $\underline{z} = (z(s_1), \ldots, z(s_n))'$ the corresponding vector of data values at observed spatial locations $\underline{s} = (s_1, \ldots, s_n)'$. Suppose we want to predict a vector of k random variables $\underline{Z}_0 = (Z(s_{0,1}), \ldots, Z(s_{0,k}))'$ at unobserved spatial locations $\underline{s}_0 = (s_{0,1}, \ldots, s_{0,k})'$. We assume that the mean function for Z and Z_0 can be written as

$$E(\underline{Z}) = \underline{\mu}(\underline{s})$$
$$E(\underline{Z}_0) = \underline{\mu}(\underline{s}_0),$$

where $\underline{\mu}(\underline{s})$ and $\underline{\mu}(\underline{s}_0)$ are $n \times 1$ and $k \times 1$ dimensional mean vectors associated with data locations $\underline{s}$ and prediction locations $\underline{s}_0$, respectively.

We define the link function

$$\underline{\eta} = g(\underline{\mu}(\underline{s})) = X\underline{\beta}, \tag{1}$$

where X is an $n \times p$ matrix of explanatory variables or a design matrix at observed locations and $\underline{\beta}$ is a $p \times 1$ vector of parameters. Alternatively the mean function may be written as

$$\underline{\mu} = E(\underline{Z}) = h(X\underline{\beta}),$$

where $h(\cdot) = g^{-1}(\cdot)$ is the inverse link function. In case of our example with the number of new start-ups, the canonical link function is the log link $\underline{\eta} = \log(\underline{\mu}(\underline{s}))$ for Poisson distributed data.

Further we assume that

$$var\begin{pmatrix}\underline{Z}\\ \underline{Z}_0\end{pmatrix} = V = \begin{bmatrix}\sum_{ZZ} \sum_{Z0}\\ \sum_{0Z} \sum_{00}\end{bmatrix},$$

where $\sum_{ZZ}$,$\sum_{Z0}$, and $\sum_{00}$ are known positive definite matrices of dimensions $n \times n$, $n \times k$ and $k \times k$, respectively. In practice the general symmetric positive definite variance-covariance matrix V can be calculated by

$$V = \text{var}(\underline{Z}) = v_{\underline{\mu}}^{1/2} R(\underline{\alpha}) v_{\underline{\mu}}^{1/2}, \tag{2}$$

where $v_{\underline{\mu}} = diag[v(\mu_i)]$, and $v(\mu_i)$ is the general form of the variance function. The matrix R is the correlation matrix that describes the spatial dependence among the observations, which in practice will be estimated by a semivariogram model with parameter vector $\underline{\alpha}$, denoting the parameters nugget effect, partial sill, range. In terms of mátern semivariogram model, introduced by Handcock and Stein [6], $\underline{\alpha}$ will be extended by the smoothness parameter. For Possion data $v(\mu_i) = \mu_i$ and the variance-covariance matrix V in (2) can be written as

$$V = diag[\mu_i]^{1/2} R(\underline{\alpha}) diag[\mu_i]^{1/2}.$$

Gotway and Stroup [5] emphasize that estimation with the generalized linear model is a maximum likelihood procedure, but that the full log-likelihood for estimating $\underline{\beta}$ is not needed. It is sufficient to describe the relationship between the mean and the model with the link function, the form of the variance and the relationship between the variance and the mean. This is fulfilled by the quasi-likelihood procedure (cp. [5,9: 323ff]).

Prediction with generalized linear models can be accomplished by obtaining $\hat{\underline{\beta}}_G$ as an iterative solution vector corresponding to equation

$$X'WX\underline{\beta} = X'W\underline{z}^*,$$

where $W = D'V^{-1}D$, $D = diag[\partial\mu_i/\partial\eta_i]$ is an $n \times n$ matrix, and $\underline{z}^* = \underline{\eta} + D^{-1}(\underline{z} - \underline{\mu})$. Further estimating $\underline{\eta}$ as $\hat{\underline{\eta}} = X\hat{\underline{\beta}}_G$, corresponding to the data, and $\hat{\underline{\eta}}_0 = X_0\hat{\underline{\beta}}_G$, corresponding to the variables to be predicted. Then $\hat{\underline{\mu}}(\underline{s}) = h(\hat{\underline{\eta}})$ and $\hat{\underline{\mu}}(\underline{s}_0) = h(\hat{\underline{\eta}}_0)$ can be used with the appropriate covariance Matrix V in (2), to calculate $\hat{\underline{Z}}_0$ from

$$\hat{\underline{Z}}_0 = \hat{\underline{\mu}}(\underline{s}_0) + \Sigma_{0Z}\Sigma_{ZZ}^{-1}(\underline{Z} - \hat{\underline{\mu}}(\underline{s})),$$

Gotway and Stroup [5] emphasized that the estimated parameter vector $\hat{\underline{\beta}}_G$ will still be consistent for $\underline{\beta}$ even if the correlation matrix is not correctly specified.

A further extension of general linear model theory introduced by Hastie and Tibshirani [7] are the general additive models. The generalized additive model differs from the generalized linear model in that an additive predictor $\sum_j f_j(X_j)$ replaces the linear predictor $X\underline{\beta}$ in (1). This theory should be adapted to apply it with spatial data.

3.2 Spatial Bayesian Model with Count Data

Ver Hoef and Frost [15] develop a Bayesian hierarchical model for analyzing trend, abundance, and effects of covariates for monitoring programs of multiple sites and apply it to counts of harbor seals in Alaska. This approach also has broader application in other monitoring situations. We show in this section, how we can apply this Bayesian hierarchical model for analyzing trend and effects of covariates for monitoring the number of new start-ups.

Let Z_{ij} be a random variable of the numbers of new firm foundations in the j-th year and in the i-th region. The Bayesian hierarchical model in Ver Hoef and Frost [15] begins with Poisson regression for each observation. We write

$$f(z_{ij}) = \exp(-\lambda_{ij})\lambda_{ij}^{z_{ij}}/z_{ij}!,$$

with

$$\ln(\lambda_{ij}) = \theta_{ij} + \underline{x}'\underline{\beta} + \epsilon_{ij},$$

where θ_{ij} is an intercept, $\underline{x} = (x_{1ij}, \ldots, x_{pij})'$ is a $p \times 1$ vector of observed values of covariates in region i and year j, $\underline{\beta} = (\beta_{1i}, \ldots, \beta_{pi})'$ is a $p \times 1$ vector of parameters, and ϵ_{ij} is an overdispersion parameter. We assume that conditional on the covariates, all observations are independent, then we can write the joint density

$$f(\underline{z}|\underline{\theta}, \underline{\beta}) \equiv \prod f(z_{ij}).$$

Further a separate trend model for each region is developed. In Ver Hoef and Frost [15] $f(\theta_{ij}|\tau_i, \delta^2) = N(\tau_i, \delta^2)$ is a normal distribution and the joint distribution can be written as

$$f(\underline{\theta}|\underline{\tau}, \delta^2) = \prod_i \prod_j f(\theta_{ij}|\tau_i, \delta^2).$$

In the next level of hierachy the region specific covariate parameters are grouped. The joint distribution is given by

$$f(\underline{\beta}|\underline{\mu}, \underline{\sigma}) = \prod_p \prod_i f(\beta_{pi}|\mu_p, \sigma_p^2),$$

where in Ver Hoef and Frost [15] the region specific covariate parameters are given a normal distribution with mean μ_p and variance σ_p^2.

Further in this level of hierachy the region specific covariate parameters for the trend parameters are grouped. Ver Hoef and Frost give them a normal distribution with mean η and varince γ^2. The joint distribution is given by

$$f(\underline{\tau}|\underline{\eta}, \underline{\gamma}) = \prod_i f(\tau_i|\eta, \gamma^2).$$

The overdispersion parameters are grouped and the joint distribution is

$$f(\underline{\epsilon}|\underline{0}, \underline{\xi}) = \prod_i \prod_j f(\epsilon_{ij}|0, \xi_i^2),$$

and

$$f(\underline{\xi}|\nu_a, \nu_b) = \prod_i f(\xi_i|\nu_a, \nu_b).$$

In Ver Hoef and Frost [15] $f(\epsilon_{ij}|0,\xi_i^2)$ is a normal distribution with mean 0 and variance ξ_i^2 and $f(\xi_i|\nu_a,\nu_b)$ is a gamma distribution with parameters ν_a and ν_b.

In the fourth and final level of the hierarchy diffuse priors have to be given to μ_p, σ_p^2, δ^2, η_q, γ_q^2, ν_a and ν_b. Ver Hoef and Frost [15] give the mean parameters μ_p and η_q a normal distribution with mean 0 and variance 1,000,000 to constitute the uncertainty. A gamma distribution with parameter a and b equal 0.001 is given to σ_p^2, δ^2, γ_q^2, ν_a and ν_b.

With the Bayes theorem we can write the posterior distribution

$$
\begin{aligned}
&f(\underline{\theta},\underline{\beta},\underline{\tau},\underline{\epsilon},\delta^2,\underline{\xi},\underline{\mu},\underline{\sigma},\underline{\eta},\underline{\gamma},\nu_a,\nu_b|\underline{z}) \propto \\
&f(\underline{z}|\underline{\beta},\underline{\theta})f(\underline{\beta}|\underline{\mu},\underline{\sigma})f(\underline{\theta}|\underline{\tau},\delta^2)f(\underline{\epsilon}|\underline{0},\underline{\xi})f(\underline{\tau}|\underline{\eta},\underline{\gamma})f(\underline{\xi}|\nu_a,\nu_b) \\
&f(\delta^2)f(\underline{\mu})f(\underline{\sigma})f(\underline{\eta})f(\underline{\gamma})f(\nu_a)f(\nu_b).
\end{aligned}
$$

Markov Chain Monte Carlo technique (MCMC) make it possible to obtain samples from posterior distribution. From these samples functions and summaries of the posteriori distribution can be computed. For an alternative method to obtain the predictive density see Pilz and Spoeck [10].

4 Conclusion

In this paper we have briefly discussed some ideas for presenting entrepreneurship data and have pointed out how sensible the right choice of a suitable measure can be, to compare regions or political districts. Further we have briefly presented two approaches for spatial modeling entrepreneurship data, the generalized linear model from Gotway and Stroup [5] and the Bayesian hierarchical model for count data from Ver Hoef and Frost [15]. In a future project we will test how good these two approaches can be applied to model firm foundations and firm survival in respect to their spatial location. A main problem in the latter approach will be to find suitable prior distributions for the parameters and to calculate the posterior distribution with Markov Chain Monte Carlo techniques.

References

1. Breitenecker R (2007) Räumliche Lineare Modelle und Autokorrelationsstrukturen in der Gründungsstatistik - Methodische Analyse und empirische Tests. PhD thesis, University of Klagenfurt
2. Egeln J, Gassler H, Schmidt P (1999) Regionale Aspekte von Unternehmensneugründungen in Österreich. Nomos-Verlag, Baden-Baden
3. Fritsch M, Grotz R, Brixy U, Niese M, Otto A (2002) Gründungen in Deutschland: Datenquellen, Niveau und räumlich-sektorale Struktur. In: Schmude J und Leiner R (eds) Unternehmensgründungen: Interdiszipiläre Beiträge zum Entrepreneurship Research. Physica-Verlag, Heidelberg: 1–31
4. Fritsch M, Niese M (1999) Betriebsgründungen in den westdeutschen Raumordnungsregionen von 1983–97. Freiberg Working Papers, No. 20
5. Gotway CA, Stroup WW (1997) A Generalized Linear Model Approach to Spatial Data Analysis and Prediction. *Journal of Agricultural, Biological, and Environmental Statistics* Vol. 2, No. 2: 157–178
6. Handcock MS, Stein ML (1993) A Bayesian Analysis of Kriging. *Technometrics*, Vol. 35, No. 4: 403–410
7. Hastie TJ, Tibshirani RJ (1990) Generalized Additive Models. Chapman & Hall, London
8. Koboltschnig RG (1998) Anwendungen Bayesscher Modelle in der Räumlichen Epidemiologie am Beispiel von Lungen-Karzenomdaten in Westösterreich. PhD thesis, University of Klagenfurt
9. Mc Cullagh P, Nelder JA (1989) Generalized Linear Models. Second Edition, Chapman & Hall, London
10. Pilz J, Spoeck G (2008) Why do we need and how should we implement Baysian Kriging methods?, *Stochastic Environmental Research and Risk Assessment*, Vol. 22, No. 5: 621–632
11. Schwarz EJ, Grieshuber E (2003) Vom Gründungs- zum Jungunternehmen: Eine explorative Analyse. Springer, Wien
12. Silverman BW (1986) Density Estimation for Statistics and Data Analysis. Chapman & Hall, London
13. Statistik Austria (2003) Arbeitsstättenzählung 2001: vorläufige Ergebnisse. http://www.statistik.at/wdbs/jsp/aztabellen.jsp

14. Struck J (1999) Quo Vadis Gründingsstatistik?. DtA, Wissenschaftliche Reihe, Band 10, Berlin
15. Ver Hoef JM, Frost KJ (2003) A Bayesian hieracichal model for monitoring harbor seal changes in Prince William Sound, Alaska. *Environmental and Ecological Statistics*, Vol. 10: 201–219
16. Wirtschaftskammer Österreich (2003) Unternehmensgründungen in Österreich. Wien

Part III

Integrated Information Systems: Combining (Geo)Statistics, GIS and RDBMS

GIS, Users, Developers, and Spatial Statistics: On Monarchs and Their Clothing

Konstantin Krivoruchko[1] and Roger Bivand[2]

[1] Environmental Systems Research Institute Redlands, Redlands, CA, USA
kkrivoruchko@esri.com

[2] Norges Handelshøyskole, Bergen, Norway
Roger.Bivand@nhh.no

1 Introduction and Motivation

The development and documentation of software for the analysis of geographical data is maturing, and the needs and desires of varying user communities are becoming clearer[3]. Certainly today there are more users in more communities, and in general much more data than before, even though data is more accessible in some countries than in others. Many more users are now meeting geographical data through geographical information systems software (GIS). GIS are general-purpose environments for handling geographical data, and do not assume that the user will need to make predictions or draw inferences from the data, or error propagation in geographical data analysis. Indeed, much of current progress in GIS is in making it easier for users to construct maps at the front end and in providing open and consistent data base support at the back end. Neither of these two areas lie close to the central concerns of statistical data analysts, such as making predictions with associated uncertainties, but can be of great value to them.

In meeting and undertaking dialogues with users and developers, it seems both valid and important to attempt to explore some of the assumptions the different communities hold themselves, have about each other, and the tasks they undertake separately and jointly. Some of the points to be made will draw on the ontology discourse in geographical information science (GIScience), which may be helpful in throwing light on different assumptions made by different communities, not just technical/motivational, but also related to the sociology of organizations and of scientific disciplines.

The paper discusses these issues in general terms, but more specifically touching on tools and methods that may propagate between communities of users, and on difficulties associated with the use of inference in inappropriate

[3] This paper represents the views of the authors as they were when it was written in 2003.

settings. In particular, we will present and discuss selected examples of analytical practice that are common, although alternatives exist that resolve or avoid some of the difficulties of these methods. In some cases, choices were limited at the time the methods were proposed by lack of access to computing power and capacity, in others by lack of access to appropriate software. In further cases, choices of methods seem to reflect barriers to the diffusion of practise across discipline boundaries, in particular from the statistical sciences, especially applied statistics, to other fields. This also reflects some of the organizational relationships, in that for example work in one field is most frequently refereed within that field, so that duplications may not be made plain for some time if at all. On the other hand, scientific progress in ever-more specialized fields is difficult to track, so multiple apparently original work is more the rule than the exception, and indeed should be welcomed as providing replicated indications of the potential fruitfulness of an approach. It does however introduce the risk of "borrowing" between different fields of applied science without sufficient reference being made, and/or without an adequate understanding of the underlying assumptions. Analogies can be useful, but can also be misleading, because the history and rationale of the development of a method may be discarded when it is "transplanted" into a new field. The transfer and rephrasing of ideas about geostatistics from meteorology to geology is a classic example: compare [14, 6, 7, 8, 20, 21].

We will also examine needs for documenting analyses leading to decisions, because decision support involves not only giving policy advice, but also providing a clear statement of how the advice was reached. This means not only documenting data sources and collection methods, but also how the data has been handled and analyzed. Analysis is partly a matter of the formal definitions of methods, but should specify the implementation or implementations used to create derivative products used as a basis for policy advice. This means that it should be possible, as in food and drugs appraisal procedures, to assign responsibility for each step in data analysis, so that another researcher could replicate it and confirm that the results achieved are in accord with the data and the specified analytical scheme. This is analogous to the statement of authority in metadata sources. For statistical analysis, this is known as reproducible statistical research.

This leads on to a discussion of the benefits different software implementations can offer one another, be they closed or open source. While open source software by its nature provides full insight into algorithm implementation, it may be no better or worse than other implementations in providing facilities for recording the steps taken in data analysis. It should also be acknowledged that different user communities expect and require different levels of support and documentation, and that developers should have some grasp of their needs for and use of implemented methods. This may be balanced by developers offering guidance, either narrowly prescriptive: do this with data of that class, or broadly prescriptive: with data of that class, consider the assumptions you are making, and try at least some of the following alternative methods to check the robustness of your conclusions.

Before proceeding, the reader deserves an explanation for the extension of the title we have chosen. We found it helpful to stress that even when an analyst feels "well-clothed" in relation to the expectations of his or her community, it may well be that others will have different opinions. This is about assumptions of what "well-clothed" means in different circumstances, and about the sometimes self-reinforcing views expressed about this in inward-looking communities of both users and developers. Different users and research communities do things in various ways, often tradition-based, and it is a user's inherent right to choose software and tools to use in his/her research and decision-making. But this right requires that the user (or developer making tools available) accepts responsibility at least for documenting how the analysis has been done. It is not enough to rely on the assurances of courtiers that we are fashionably clothed, when our apparel is awry or absent in the view of others.

2 Assumptions Held by User and Developer Communities

The landscape in which we are living and working is a reality that is much more complicated than the statistical framework in most studies. It is definitely not a flat surface, see Fig. 1. There is nothing to measure using simple straight-line distances. There are many natural and artificial barriers between spatial objects. Among these are disciplinary barriers, which make it difficult to interact fruitfully with people from other disciplines and traditions. The naïve GIS user is shocked at how different reality is from the models that describe it. There is no such thing in Nature as the Gaussian distribution and data homogeneity. There are no clear boundaries between polygonal objects. Both GIS users and developers are users of reality, but they may approach its conceptualization differently.

In the real world users of GIS and spatial statistical software vary in their insight both into statistics and into their own discipline-based domains in depth and breadth. Some users do not have either the programming experience or the motivation to modify and customize the software to suit the needs of analysis, while others will want to do so. Kuan [18] terms this the integration of the consumer into production, and this is applicable not only to software development, but to many kinds of scientific endeavors. An immediate consequence is that if we really want to make Spatial Statistics available to the average GIS user, which includes tightly integrated spatial statistics in a GIS environment and "protecting" the user from themselves (that is, inappropriate use of methods), then an attractive approach is what is already done by ESRI, building Geostatistical Analyst within ArcMap. There are some problems with such an approach for users who feel themselves well educated

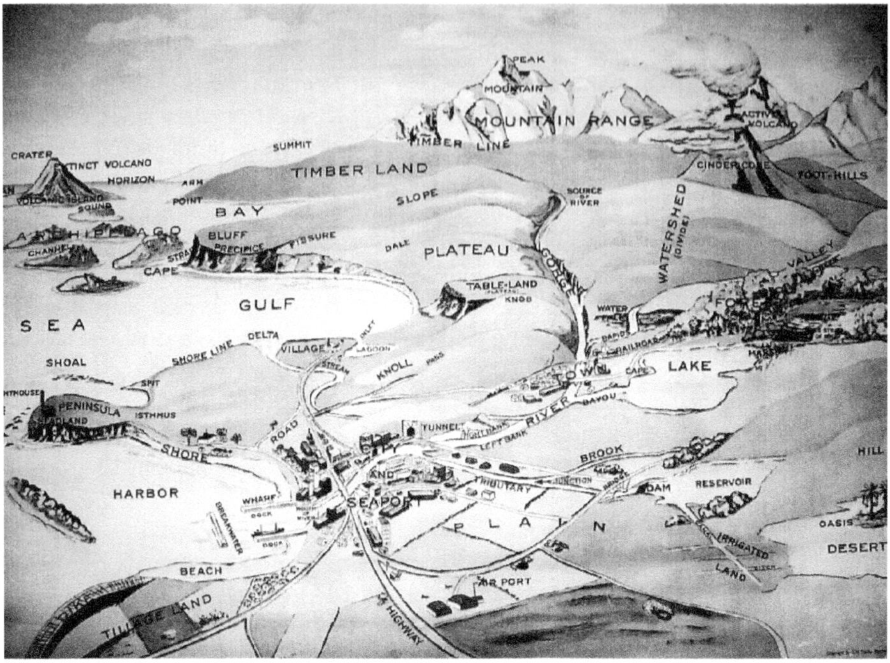

Fig. 1. Typically the landscape in which we are living and working is much more complicated than our models assume

in statistics/geostatistics and thus able and willing to use additional methods or modify existing ones; we will discuss one possibility to resolve existing problems below.

The researcher as user and as developer is often stressed by conflicting understandings of why analysis is being undertaken and what kinds of results are acceptable, and to whom. In academic settings, results are scanned for propriety by referees, using the standards of their communities. Curiously, they quite often disagree, either on standards, or on how they "read" the product with which they have been presented. In applied settings, reviewing and evaluation is also practiced, but there are most often also instrumental goals for the analysis being carried out. These typically do not adequately acknowledge the fact that uncertainty is certain. This is a key point of discord between the use of spatial statistical methods, in which we need both predictions and estimates of uncertainty, and the demands of practical users, or rather the organizations for which they work, for certainty.

There are as many logics as we can imagine. This is because logic is based on systems of axioms and rules for deriving logically true statements. For example, the greatest mathematician of the 20th century, Andrei Kolmogorov, formulated the following rule of human logic: Let $[P \Rightarrow Q]$ and $[Q$ is nice$]$; then $[P]$. Example: If my parents have money, I'll have a new bicycle $[P \Rightarrow Q]$;

It is nice to have a new bicycle [Q is nice]; Hence my parents have money [P]. In fact, many users follow such logic, but there are other possibilities.

Chrisman [3] points to weaknesses in the use of ontology-derived terms, such as *bona fide* and *fiat* objects, arguing that the distinction between them is not very helpful in practice: "The notion of *fiat* objects does recognize the human dimensions of practice, but some of these same issues recur in the objects that are meant to be beyond human interference. *Bona fide* objects are just as subject to conventions and standards developed from disciplinary practice." This observation has direct application to spatial statistics, in that disciplinary practices mediate in the treatment of objects, especially when the observed data objects have been used as a basis, with selected methods, for modeling and prediction.

One specific difficulty is that spatial statistics as a field is broader than the application of these methods to geographical data. Some methods, in particular point pattern analysis, but also others, have direct relevance at much larger and smaller scales, such as in the analysis of patterns on microscope slides, or in medical imaging. Because of this, the views of reality embodied in methods of analysis may or may not be well suited to geographical data. Other views of reality within geographical scales are difficult to represent using legacy GIS data models, for example time and a third dimension.

It is arguably the case that even good statistical training will not help someone who is lacking in domain knowledge terms, so that some "positive" or "self-reinforcing" intersection of methods practice and domain knowledge is desirable. If a researcher does not understand his own discipline, methods or software will not help, but well-structured methods and software can "enable" scientists who are aware of the often difficult assumptions of their own domain. Some of the methods may actually be simple, just good practice in data analysis, like good laboratory practice. This is associated with learning, engaging users by asking them questions to try to get them to grasp and operationalise their research problem in a way that lets them both solve that problem (or admit that it cannot be solved as posed), and learn something more generic that is transferable to other situations encountered later.

In a similar way, user and developer communities functioning in relation to spatial statistics software, especially but not only extensible software, allow participants to become more familiar with each others' assumptions. Learning is here the key, conditioned by the willingness of participants to make their positions explicit in understandable terms. Some software includes methods and implementations that would not be proposed by statisticians, but are provided because the domain scientists expect them to be present, if just because that were once considered appropriate, and were fashionable when the scientists were trained. It is also important to acknowledge that statistical methods are often seen by domain scientists as not very enjoyable, compared for example with fieldwork. Promoting positive attitudes towards analysis ought to be included in the development of such software, and is in many cases neglected, because the developer does not share this dislike. If developers were forced to do fieldwork instead of coding, they might empathize better with users of their software. The message needs to be sent clearly that the users' data, collected with considerable commitment and interest, and often expense, deserve the respect of the developer, embodied in the software – executables, documentation, and training, and in the virtual community (online discussions, meetings, conferences). Advances in computing hardware may lead to another problem: the use of ever more popular MCMC methods risk "dumbing down" science, making statistics mindless computing in cases where compute power is used instead of analysis.

3 Selected Examples of Conceptual Discourses and Discords

We will present one extended example of discord between practices in spatial statistics, concerning indicator kriging usage. The presentation of probability values for local indicators of spatial association is another topic that will be mentioned more briefly.

3.1 Indicator Kriging Usage

Journel proposed indicator kriging [13] as an alternative to disjunctive and multiGaussian (that is kriging after data transformation) kriging, because they require good understanding of the assumptions involved and were considered difficult to use at that time. In indicator kriging the data are preprocessed first. Indicator values are defined for each data location as the following: an indicator is set to zero if the data value at the location s is below the threshold, and one otherwise:

$$I(s) = I(Z(s) < \text{threshold}) = \begin{cases} 0 \; Z(s) < \text{threshold} \\ 1 \; Z(s) > \text{threshold} \end{cases}$$

Indicator transformation for one-dimensional data is illustrated in Fig. 2. Transformed input data inside the interval around threshold (for example,

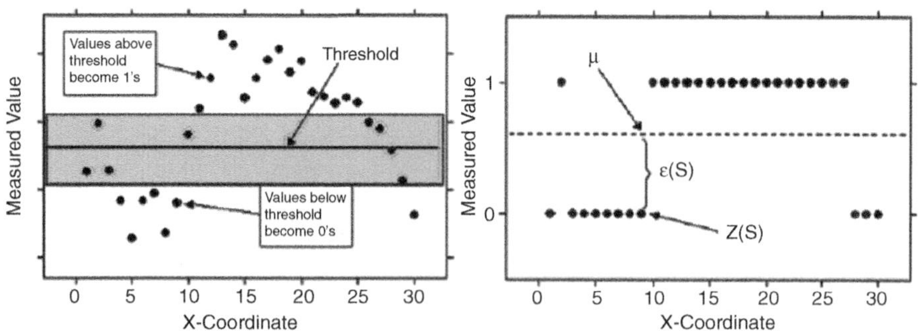

Fig. 2. Illustration of the indicator transformation for one-dimensional data

this can be measurement error interval) are displayed as red points in the indicator transformation in the graph to the right. It is quite possible that they can be exchanged if one more measurement will be taken.

Notice that after indicator transformation, input values near and far from the threshold became either zeroes or ones. This means that we are loosing information when transforming the data.

Then these indicator values are used as input to ordinary (sometimes to simple or universal) kriging. Ordinary kriging produces continuous prediction and we might expect that prediction at the unsampled locations will be between zero and one (this is often not fulfilled in practice, however). The prediction is interpreted as the probability that the threshold is exceeded at location s. For instance, if the prediction equals 0.71, it is interpreted as a 71% chance that threshold was exceeded. Predictions made at each location form a surface that can be interpreted as a probability map that the specified threshold is exceeded. If a *set* of indicators is used as input to ordinary kriging (for example, 10 quantiles of the input data distribution), the resulting set of predictions at each location can be combined to give a cumulative probability distribution from which a probability density distribution can be estimated and the prediction mean and variance can be calculated.

Although indicator kriging became very popular immediately, a number of problems have been found. If different semivariogram models are used for different thresholds, then internally inconsistent results may be obtained. One possible workaround for this problem is to use the median indicator variogram for all indicators. However, this nullifies the potential advantage of the model that the spatial structure of a variable depends on its value. For instance, we might expect that range of correlation is smaller and variance is larger for large values. Nowadays indicator kriging is mostly used to provide risk-qualified predictions (probability that a specified threshold is exceeded) at the unsampled locations, and not for prediction itself.

Consider this kriging model for the signal $Y(s)$, (see [17]):

$$Y(s) = m(s) + S(s) + \eta(s)$$

where $m(s)$ is a large scale variation (trend), known or estimated, $S(s)$ is a random process with zero mean and known covariance (small scale variation), and micro-scale variation $\eta(s)$ is the variation at a scale, too fine to be recognizable from the data. Measurement Z_i in the location s_i is a sum of the signal and independent random error with zero mean and known variance.

$$Z_i = Y(s_i) + \varepsilon_i, i = \overline{1, n},$$

where n is a number of measurements. This allows for more than one measurement at the same data location.

Geostatistical prediction and conditional simulations should not honor the data if there is measurement error and all real data are not exact. But geostatistical programs usually assume that data are perfect, that is $\varepsilon_i = 0$, which contradicts common sense. Probably this comes from a distrust of statistical models. Users think that their data are all they really know, see discussion in [17].

The idea behind indicator kriging is to estimate probability that the specified threshold T was exceeded:

$$Pr(Y(s) \geq T|Z) = E(I(s) \geq T|Z)$$

assuming that there is no measurement error, that is $Z(s) \equiv Y(s)$. A reason for this assumption is that

$$\begin{aligned} E(I(Z(s_0)) \geq T) &= Pr(Z(s_0) \geq T) \\ &= Pr(Y(s_0) + \varepsilon(s_0) \geq T) \\ &\neq E(I(Y(s_0)) \geq T), \end{aligned}$$

meaning that indicator kriging is a biased predictor of a signal and this bias can be substantial if measurement error is large. The non-existence of the filtered predictant is a serious disadvantage of the indicator kriging model. In practice, predictions in the close vicinity of the data locations are usually not close to 0 or 1 and predictions jump to 0 or 1 at the data locations. Such a prediction surface for discrete input data must be questioned.

Kriging is the best linear predictor for Gaussian random variables, but $I(Z(s_i)) \geq T)$ are Bernoulli random variables and the indicator predictor may be far from optimal [5]. A semivariogram might be inappropriate measure of spatial continuity of discrete data, [25].

There is also a problem with block estimation of $I(Z(B)) \geq T)$, where B is an area, from point data Z_i since

$$I(Z(B)) \geq T) = I\left(\frac{1}{|B|}\int_B Z(u)du \geq T\right) \neq \frac{1}{|B|}\int_B I(Z(u) \geq T)du$$

see discussion in [11], meaning that an additional assumption concerning the covariance between point and block indicators, $\mathrm{cov}(I(Z(s_i)) \geq T), I(Z(B)) \geq T))$, needs to be made.

A basic assumption behind any standard geostatistical model is an assumption about data stationarity. In reality, data often more or less depart from stationarity, and the solution is to use detrending and transformation techniques to make data close to stationarity, see case study with comparison of indicator, disjunctive, and other krigings performance in Krivoruchko [15]. However, indicator kriging uses original data and there is no possibility to transform data to stationarity. Also, even if the original measurements are stationary, there is no guarantee that the transformed indicator variable will be stationary. For simple statistical models, departures from the stationarity assumption are more serious in their consequences for the reliability of inference than violation of the distribution assumption for more complex models.

An important advantage of statistical models over deterministic ones is the possibility to estimate prediction uncertainty. Without data pre-processing,

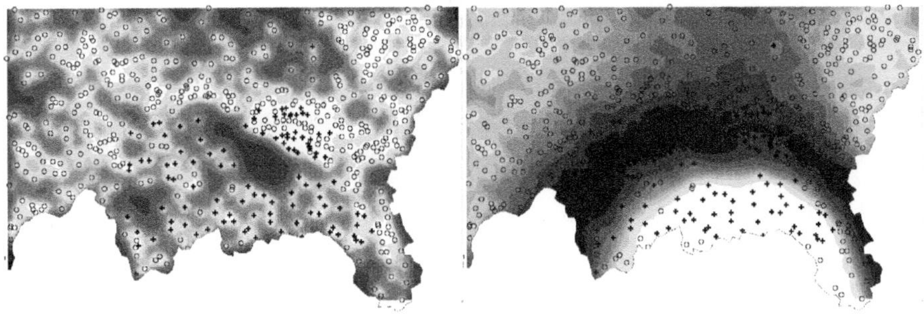

Fig. 3. Comparison of standard error of indicators maps created using indicator (*left*) and disjunctive kriging (*right*) after data transformation using threshold 15Ci/km^2

the kriging standard error map does not depend on data values, only on measurement density. If input data are transformed to an approximately Gaussian distribution, prediction standard errors depend on data values. For example, Fig. 3 taken from [15], compares standard error of indicators maps created using indicator and disjunctive kriging with data transformation for radionuclide soil contamination interpolation in Southern Belarus. The probability map created using disjunctive kriging is data dependent, and the largest uncertainty corresponds to areas close to the selected threshold value. Without reliable information on modeling uncertainty, decision-making may be misleading.

This example shows how the practical use of methods can become encumbered with what we can call "encrustations". A method became established, that was introduced to address a pragmatic issue, or a group of issues, at least partly because other methods, acknowledged to be more adequate, were seen as practically or computationally infeasible, as well as poorly matched to users' possibilities. Over the intervening period, not only computational resources, but also the research bases, have changed, but not least for pragmatic reasons, analytical practice has not necessarily followed up. We could have chosen to present other examples of areas where spatial statisticians differ sincerely in their approaches to analysis, and others have been noted above in brief. This will for now have to be sufficient to indicate some of the features of one of many debates.

Because of the problems described above regarding indicator kriging usage, it is safe to use it as ESDA technique, but not as a prediction model for decision-making.

It is not advisable to use the conditional indicator simulation model as well, because of above-mentioned problems with indicator kriging and because of some other problems, see Gotway and Rutherford [10].

3.2 Local Indicators of Spatial Association

The main difference between geostatistical and polygonal data analysis is in the order of specifying covariance/semivariogram matrix and weights of neighbors involved in spatial prediction. In geostatistics, the correlation between locations separated by a specified distance is modeled first. Then weights are calculated automatically. In polygonal data analysis weights are defined first. It is supposed that they reflect the statistical distance between polygons. Cliff and Ord ([4], p. 11–13) provide the initial formalization of the relationships as a generalized weighting matrix, most usually termed **W**. It is usual in the literature to define the contiguity relation in terms of sets of neighbors of zone or site i. These are coded in the form of a weights matrix **W**, with a zero diagonal, and the off-diagonal non-zero elements often scaled to sum to unity in each row, with typical elements:

$$w_{ij} = \frac{c_{ij}}{\sum\limits_{j=1}^{N} c_{ij}}$$

where $c_{ij} = 1$ if i is linked to j and $c_{ij} = 0$ otherwise. This implies no use of other information than that of neighborhood set membership. In practice set membership is almost always defined arbitrary and the most popular way to define it is on the basis of shared boundaries, centroids lying within distance bands, and "rook" or "queen" rules, terms borrowed from chess. Often it is unclear how rook or queen will behave near the boundary of the area under investigation.

Spatial autocorrelation is the term given to a measure of the correlation among neighboring values. There are many different ways to quantify spatial autocorrelation, but the most common index for regional data is Moran's I [22, 4]:

$$I_{\text{global}} = \frac{N}{\sum_{i \neq j}^{N} \sum_{j=1}^{N} w_{ij}} \frac{\sum_{i \neq j}^{N} \sum_{j=1}^{N} w_{ij}(r_i - \bar{r})(r_j - \bar{r})}{\sum_{i=1}^{N} (r_i - \bar{r})^2}$$

where $\bar{r}$ is the global mean value, defined and calculated as a simple average value based on all the data. The data could be counts or rates, although in our opinion working with count data is misleading since the underlying population also varies among the regions, see detailed discussion in [16].

A local version, called a Local Indicator of Spatial Association or LISA by Anselin [1] is:

$$I_{i,std}^{\text{local}} = \left(\frac{r_i - \bar{r}}{s}\right) \sum_{j=1}^{N} w_{ij} \left(\frac{r_j - \bar{r}}{s}\right)$$

where $\bar{r}$ and s are the overall mean and standard deviation, respectively, and the weights w_{ij} reflect the spatial proximity between regions i and j.

This statistic provides a measure of local similarity (or dissimilarity) for each region. There are problems with LISA. For instance, it is hard to understand how to interpret a case where, when using adjacency weights, two adjacent regions have very different statistics.

Getis and Ord [9] and Anselin [1] give expectations and variances for the local indicators, using both assumptions of normality and randomization, following Cliff and Ord [4] for the global measures. The standard route to drawing inferences has been to treat the square root of the difference between the observed measure and its expectation divided by its variance, as a standard normal deviate. The Gaussian distribution can be a good model for continuous data, but count data are inherently discrete. In randomization, assuming the observed values are exchangeable, the assumption of stationarity is actually made and this is violated by counts and rates in any case, and also when stationarity is not present. Other indices that allow the mean and variance of the data to vary with the population in each region, and are thus more suitable for measuring clustering in regional populations are available, see discussion in [16].

Moran's I can be modified to relax the assumption of constant mean and variance. One such statistic for rates, see Walter [26], is:

$$I_i^{\mathrm{WM}} = \left(\frac{y_i - \bar{r}n_i}{\sqrt{\bar{r}n_i}} \right) \sum_{j=1}^{N} w_{ij} \left(\frac{y_j - \bar{r}n_j}{\sqrt{\bar{r}n_j}} \right)$$

assuming that the underlying risk $\bar{r}$ to be constant over all regions and estimated from the data over the entire region. This statistic is based on properties of the Poisson distribution assuming that $E(Yi) = \bar{r}n_i$. The p-values can be computed using Monte Carlo simulation as follows [16]:

1. Generate simulated values for each region, under the null (or default) hypothesis of spatial independence; Here we assume the data follow a Poisson distribution with mean $E(Yi) = \bar{r}n_i$; these values are simulated from the Poisson distribution and are not a permutation of the observed counts.
2. Compute the statistic of interest, in this case $U = I_i^{\mathrm{WM}}$ for each simulated data set.
3. Repeat M times. This gives $U_1, U_2, \ldots, U_M$.
4. Compare the observed statistic calculated from the available data, say U_{obs} to the distribution of the simulated U_j and determine the proportion of simulated U_j values that are greater than U_{obs}

The idea is to obtain the proportion of simulated values that are more extreme that the value determined from the data.

It is natural and users regularly ask for probability values to be made available for global and local indices of spatial association, but it is a rather delicate procedure. Some software permutes all the data values across the set of units, as is typically done for global measures. But this does not provide an adequate basis for inferring about the local neighborhood, in which the range of values found may be much more restricted. One could attempt to simulate for each neighborhood, but because the numbers of neighbors are small, very few draws can be made before all possible combinations have been exhausted. An underlying problem is that global autocorrelation, perhaps reflecting a trend in the data, will yield apparently significant local measures, and will also make the use of the whole pool of data values for simulation wrong, because within a local neighborhood, the trend limits values to a narrow band. Users are at risk of drawing conclusions from the output of local indicators that are not robust, and it is not obvious how to indicate to them how dependent these indicators, and derived measures, such as probability values, are on the assumptions being made.

Often software users and developers assume that the data are independent and follow a stationary Gaussian distribution, but this is an unreliable conditions in practice, at least for aggregated data, such as cancer and crime rates. In fact almost all polygonal data are not continuous and Moran's I should arguably be used only for pedagogical purposes. The best approach is to use Monte Carlo testing. In this approach we generate realizations from a specified univariate distribution that describes the data, calculate local index for each polygon and then compute the p-value as in example using Walter's modified I above. This certainly is best approach in the case of global statistics, but it still may be misleading for LISA because number of neighbors is usually small, less than 10, and any statistics might be insufficient. One possible solution to the problem is to use several different indices as in the case study by Krivoruchko et al. [16]. If all or most of indices give similar results, we can safely make conclusion about data clustering or cross-correlation. If not, further research is required.

4 Reproducible Research

Leisch and Rossini [19] present arguments for making statistical research reproducible, so that given the same data, another analyst will be able to recreate the research outcome used in a paper or report. If software has been used, the implementation of the method applied should also be documented. If arguments used by the implemented functions can take different values, then these also need documentation. An example is the way in which a geostatistical layer in ESRI ArcMap is defined [12]. Most ArcMap layer types store the reference to the data source, the symbology for displaying the layer, and other defining characteristics. A geostatistical layer stores the sources of the data from which it was created (usually a point feature layers), the symbology, and other defining characteristics, but it also stores the model parameters from the interpolation, including type or model for data transformation, covariance and cross-covariance models, estimated measurement error, trend surface, searching neighborhood, and results of validation and cross-validation.

Another example of model storage for reproduction and updating is the geoprocessing environment, see [2]. Geoprocessing tools allow researchers to combine and interpret data obtained from different sources. As such, they form important components of an underlying model that takes input data (coverages, shape files, raster grids) and assimilates them in a meaningful way to produce output information that can provide a more suitable interpretation.

The ease with which geoprocessing can be implemented within a GIS makes it easy to forget that such processes inherently alter the input data. In most cases, the geometric properties of the features of the input data are altered to form new features and functions of the input attribute values are transferred to the new features. In many GIS applications, geoprocessing is just a means to an end. In others, however, a more thorough understanding of the model may be desirable. The overall goal of modeling may be to understand how assumptions, parameters and variation associated with the input data affect the resulting output data and the conclusions made from them. In such cases, a probabilistic framework for model building with geoprocessing tools may be desirable. Consider the following geoprocessing scenarios:

- A buffer function is used to create a zone of a specified distance around the features in a layer. How do we know what distance to use? What happens to our results and conclusions if we increase the distance slightly? Given a choice of distance, how wrong can our conclusions be? This latter question can have huge implications in environmental justice, for instance, where we are trying to decide if under-privileged people are more likely to live near toxic waste sites, landfills, or other environmental hazards.
- Data are usually available at different resolutions. Union and intersection geoprocessing operations require that data be aggregated and disaggregated. For example, a soil classification map with polygonal features is often converted to raster for use in geoprocessing; DEM data exist in just several resolutions and are often upscaled or downscaled to provide elevation estimates needed for geological and hydrological applications; spatial interpolation methods such as kriging and inverse distance-squared are often used to provide maps of environmental variables whose values are then aggregated to in order to link them to public health and disease information summarized for administrative regions. However, raster conversion, interpolation, and aggregation and disaggregation procedures provide only estimates for attributes associated with the newly created features, not the true values. The user may need to know the accuracy of the resulting estimates, and the impact of estimation error on additional calculations.
- The accuracy of spatial data is a very important concern. Locational (positional) errors occur when the geographical coordinates of a point feature are not known precisely. This can be due to measurement error, projection distortion, or reporting errors. Even if locational error is relatively small, the uncertainty arising from lack of precise geographical coordinates can

have a substantial impact on spatial analysis. For example, spatial proximity is usually calculated using distances between pairs of data locations and uncertainty in location coordinates will influence the results of the raster-based interpolation methods in geostatistical analyses. When these output raster layers are used as input to other geoprocessing operations, the errors propagate through the calculations and small errors can quickly add up to large errors if many calculations are performed. If uncertain locations are buffered, attribute values from another layer may be misclassified. GIS users may want to make sure that locational errors do not greatly impact their results and conclusions, and if they do, they might want to be able to track them or adjust their results for them.

In all of these cases it is important to maintain the lineage of the operations being performed, so that a trail exists allowing either an audit to re-create the research, or to permit additional examination of changes in conclusions when data or function arguments are supplemented or modified. Introducing error propagation in geoprocessing operations may lead to irreproducible results (layers and maps), but nearly reproducible research: if result is a prediction with associated uncertainty, it is a realization of the true process, which inherently unknown, hence, irreproducible.

An example of a setting in which the user may wish to supplement the geoprocessing model is when the output of a function or procedure is in a form that is harder to display in map or tabular form. Say that we have a map layer with point locations of some events. They appear clustered in some sense, and we can construct and plot maps of the density of the pattern using different bandwidths. But we would like to test whether the spatial pattern of points could have been generated by a cluster process, within a given study area polygon. Typically, the output will be a plot or summary statistic for the pattern as a whole, and may vary depending on arguments to the function. Consider the example of a test of a point pattern representing the places of residence of juvenile offenders in a part of Cardiff, Wales.

Point pattern analysis is concerned with the location of events, and with answering questions about the distribution of those locations, specifically whether they are clustered, randomly or regularly distributed. Point pattern analysis is very sensitive to the definition of the study area, since a regularly distributed pattern can be made to seem clustered by including large margins within the study area. Measures are also subject to boundary corrections, and most often study area boundaries have to be defined as convex polygons over the study area, or in the simplest form as rectangles bounding the points under analysis. The simplest way of exploring point pattern data is by examining a two-dimensional frequency distribution of counts within equal-area units imposed on the study area, giving an impression of how the intensity of the point process varies; this can be extended to kernel estimation. Nearest neighbor distances are also used to analyze intensity of points, the mean number of events per unit area at point $\mathbf{s}$. Spatial dependence is captured by the second order properties of a spatial point process, which involve the relationship between numbers of events in pairs within the chosen study area. The K function is a summary measure of second order effects, and is estimated for a sequence of rings of distance h by:

$$\hat{K}(h) = \frac{R}{n^2}\sum\sum_{i \neq j} \frac{I_h(d_{ij})}{w_{ij}}$$

where R is the area of the study area polygon, n is the number of points, $I_h(d_{ij})$ is an indicator function which is 1 if $d_{ij} < h$ and 0 otherwise, and w_{ij} is a edge adjustment – the proportion of the circumference of a circle centered on i and going through j that is within the study area. $\hat{K}(h)$ is often reported as $\hat{L}(h)$, where:

$$\hat{L}(h) = \sqrt{\frac{\hat{K}(h)}{\pi}} - h$$

Observed values of $\hat{K}(h)$ for a given study area polygon boundary can be compared with simulated values of the same measure for a given spatial point process model. Most often the model chosen is that of complete spatial randomness, which involved simulating n points within the study area polygon for each simulated pattern following a homogeneous Poisson process. Results are displayed by recording $\hat{K}(h)$ for each simulation, and plotting the largest and smallest values for each h as a simulation envelope. If the observed $\hat{K}(h)$ leaves the envelope, this may be taken to show that it is unlikely that – for the chosen number of simulations – that the observed pattern could have been generated by the process used in the simulation. When we wish to test whether a pattern is clustered, it may be more natural to use a process model that suits this hypothesis. The Poisson cluster process involves the inclusion of a spatial clustering mechanism into the model, so that observed $\hat{K}(h)$ falling within a Poisson cluster process simulation envelope show that the observed pattern could have been generated by such a model.

The following code example run in the R statistical computing environment [23] will generate reproducible results, in this case the plot shown in Fig. 4. The function being called to generate the simulation envelope is pcp.sim(), contributed to the **splancs** package [24] by Giovanni Petris, and using faster code changing the order of calls to the random number generator contributed by Nicolas Picard, which can be turned off by setting argument vectorise.loop=FALSE.

```
    # Load the "splancs" package
library(splancs)
  # Load the Cardiff juvenile offenders domiciles point
  # data set and bounding polygon, assign the distance
  # sequence and compute Khat
data(cardiff, package="splancs")
r <- seq(2, 30, by = 2)
K.hat <- khat(as.points(cardiff), cardiff$poly, r)
  # Compute the fitted Poisson Clustering Process
pcp.fit <- pcp(as.points(cardiff), cardiff$poly, h0=30, n.int=30)
```

```
m <- npts(as.points(cardiff))/(areapl(cardiff$poly)*pcp.fit$par[2])
  # Set the random number generator seed and perform the simulation
  #to find the simulation envelope bounds
RNGkind(kind="Mersenne-Twister", normal.kind="Inversion")
set.seed(123)
K.env <- Kenv.pcp(pcp.fit$par[2], m, pcp.fit$par[1], cardiff$poly,
 nsim = 20, r = r, vectorise.loop=TRUE)
  # Create a function to convert Khat values to Lhat
Lhat <- function(x, r) sqrt(x/pi) - r
  # Apply the function to the simulation results
L.env <- lapply(K.env, Lhat, r)
  # plot the observed Lhat values
limits <- range(unlist(L.env))
plot(r, Lhat(K.hat, r), ylim = limits,
 main = "L function with simulation envelopes and average",
 type = "l", xlab = "distance", ylab = "", lwd=3)
  # Add the simulation average and envelope to the plot
lines(r, L.env$lower, lty = 5)
lines(r, L.env$upper, lty = 5)
lines(r, L.env$ave, lty = 6)
abline(h = 0)
```

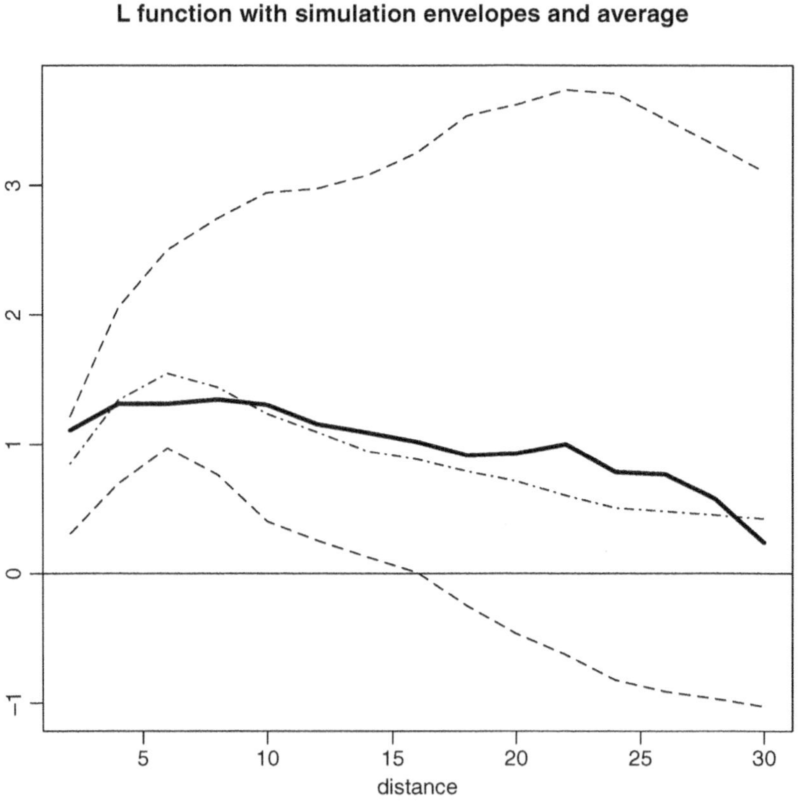

Fig. 4. Observed function for Cardiff juvenile offenders places of residence, with Poisson cluster process simulation envelope using the "Mersenne-Twister" random number generator

Running the code calculates the $\hat{K}(h)$ function from the observed spatial pattern for the chosen sequence of distances, using edge correction for a bounding polygon, and plots its $\hat{L}(h)$ transformation. We test against a Poisson cluster process by simulating such a process within the bounding box, here only 20 times, and plotting the maximum, mean, and minimum simulated $\hat{L}(h)$ values around the observed values (Fig. 4). It appears that the observed data pattern could have been generated by a Poisson cluster process.

Having the code, the specified version of the **splancs** package and R, a reviewer can re-investigate the impact on our conclusions of changing the boundaries used for calculating edge effects, the number of simulations, the distance sequence, the random number generator, and other parameters of the model. Research should be documented not just for academic reasons, and the provision of mechanisms for journaling methods used and thereby securing the lineage of objects in documents within or derived from GIS is necessary. Review and decision-making based on reproducible research using well-documented closed (Geostatistical Analyst) or open (R) code software is transparent and verifiable and does not require the participation of skilled researchers. If the steps taken are documented and can be reproduced, the results are available for checking in the future by the same or other users. Should newer methods or fresh data become available, the documentation of the lineage of results means that they can continue to be valuable for the organizations that have invested in their collection and processing.

5 Concluding Remarks

In a perfect world, we should be able to combine the strengths of statistical software and GIS, but we do not see it happening in practice except by simply passing data sets back and forth between the two. Of course, there are open source environments for both GIS and statistics, for instance, GRASS and R, but it is very unlikely that this mixture will be used for decision-making by very large community of commercial GIS software (millions of users), but only by special interest and minority groups of academics, consultants, and others needing the low-level flexibility this makes available.

Using the trivial example from the end of the previous section, we can suggest that there need be no unhealthy competition between proprietary software like ArcGIS and free software like R. Indeed, access is possible between ArcGIS and R under Windows using, for example, the R(D)COM StatConnector. R is a good candidate for testing and prototyping new statistical models for further implementation as commercial extensions for broader GIS communities. This is because it provides support for documentation and release organization, guided by a well-regarded core team. More importantly, commercial GIS software cannot be updated very often and the best way to develop and test new models is to use more flexible environment, such as growing R.

R is not polished, for example lacking an integrated GUI, and it functions more like a prototyping "kitchen", where the ingredients can be tasted before the meal is composed. The code above exemplifies this, as does access to source code for the functions at least back to the operating system libraries, if the need arises. This means that data can potentially be exchanged, in this case a set of point locations, and one or more bounding polygons, for use in R. The results could then be created as a document for display and inclusion in further work, or to provoke changes in arguments passed to the underlying R statistical compute engine.

Returning to our question about the way monarchs are clothed, we feel that there are benefits to be drawn from raising questions about the ways assumptions are handled in the statistical analysis of spatial data. In some cases, users are neither able to make nor interested in the appropriate choice of methods. In these cases, the developer should provide guidance, and document the choices made and methods used in the resulting data objects. In other cases, users are more like developers, working much more closely with the software in writing scripts and macros, and in trying out new models. Here, the accessibility of the input data objects to user-written functions is important, and for some purposes, the linking of GIS software with external statistical or modeling software may provide the level of customization some users need for their research. In both scenarios – in fact on a continuum from black-box to white-box – focus on the degree to which the assumptions of the applied methods are met will let the user, or an auditor of the user's work, find out what has been done, and hopefully avoid unnecessary blunders.

Acknowledgements

This work was supported in part by EU Contract Number Q5RS-2000-30183. The authors would like to thank Carol Gotway-Crawford for her helpful discussions.

References

1. Anselin L (1995) Local indicators of spatial association – LISA. Geographical Analysis 27:93–115
2. ArcNews (2003) `http://www.esri.com/news/arcnews/summer03articles/arcgis9.html`
3. Chrisman N (2000) Building GIS without Foundations: Ontology from a Social Practice Perspective, First International Conference on Geographic Information Science, Savannah, Georgia, USA, October 28–31, 2000, `http://www.giscience.org/GIScience2000/invited/Chrisman.pdf`
4. Cliff AD, Ord JK (1973) Spatial autocorrelation. Pion, London
5. Cressie N (1993) Aggregation in geostatistical problems. In Soares, A (ed) Geostatistics Troia 1992, vol 1, Kluwer Academic Publishers, Dordrecht, 25–36
6. Gandin LS (1959) The problem on optimal interpolation. Trudy GGO 99:67–75 (In Russian)
7. Gandin LS (1960) On optimal interpolation and extrapolation of meteorological fields. Trudy GGO 114:75–89 (In Russian)
8. Gandin LS (1963) Objective Analysis of Meteorological Fields. Gidrometeorologicheskoe Izdatel'stvo (GIMIZ), Leningrad (translated by Israel Program for Scientific Translations, Jerusalem, 1965)
9. Getis A, Ord JK (1996) Local spatial statistics: an overview. In Longley, P, Batty M (eds) Spatial analysis: modelling in a GIS environment Geoinformation International, Cambridge, 261–277
10. Gotway CA, Rutherford BM (1994) Stochastic simulation for imaging spatial uncertainty: Comparison and evaluation of available algorithms. In Armstrong M, Dowd PA (eds.) Geostatistical Simulations, Kluwer Academic, Dordrecht, 1–16.
11. Gotway CA, Young LJ, (2002) Combining incompatible spatial data. Journal of the American Statistical Association, 97:632–648
12. Johnston K, Ver Hoef J, Krivoruchko K, Lucas, N (2001) Using ArcGIS Geostatistical Analyst: GIS by ESRI, ESRI, Redlands CA
13. Journel AG, (1983) Nonparametric estimation of spatial distributions. Mathematical Geology, 15:445–468

14. Kolmogorov AN (1941) Interpolation and extrapolation of stationary random sequences. Isvestiia Akademii Nauk SSSR, Seriia Matematicheskiia 5:3–14 (Translation, 1946, Memo RM–3090–PR, Rand Corp., Santa Monica, CA)
15. Krivoruchko K (2001) Using linear and non-linear kriging interpolators to produce probability maps. Available from ESRI online at `http://www.esri.com/software/arcgis/arcgisxtensions/geostatistical/research_papers.html`
16. Krivoruchko K, Gotway C, Zhigimont A (2003) Statistical Tools for Regional Data Analysis Using GIS. ACMGIS'03, November 7-8, 2003, New Orleans, Louisiana, USA. Available from ESRI online at `http://www.esri.com/software/arcgis/arcgisxtensions/geostatistical/research_papers.html`
17. Krivoruchko K, Gribov A, and Ver Hoef J, 2000, A New Method for Handling the Nugget Effect in Kriging. Available from ESRI online at `http://www.esri.com/software/arcgis/arcgisxtensions/geostatistical/research_papers.html`
18. Kuan J (2000) Open Source Software as Consumer Integration into Production, unpublished working paper, University of California at Berkeley
19. Leisch F, Rossini A (2003) Reproducible statistical research. Chance 16(2): 46–50
20. Matheron G (1962) Traité de Géostatistique Appliquée, Tome I. Mémoires du Bureau de Recherches Géologiques et Minières, No. 14, Editions Technip, Paris
21. Matheron G (1965) Les variables régionalisées et leur estmation. Une application de la théorie des fonctions aléatoires aux Sciences de la Nature. Masson, Paris
22. Moran PAP (1950) Notes on continuous stochastic phenomena. Biometrika 37:17–23
23. R Development Core Team (2003) R: A language and environment for statistical computing. R Foundation for Statistical Computing, Vienna, Austria, `http://www.R-project.org`
24. Rowlingson B, Diggle P (1993) Splancs: spatial point pattern analysis code in S-Plus. Computers and Geosciences, 19:627–655
25. Walder O, Stoyan D (1996) On variograms in point process statistics. Biometrical Journal 38:895–905
26. Walter SD (1992) The analysis of regional patterns in health data I: distributional considerations. American Journal of Epidemiology, Supplement 132:S136–S143

Reassignment of the Farm Structure Statistical Data Using GIS and Spatialisation of the Results Based on Remotely Sensed Data

M. Sambrakos[1] and T. Tsiligiridis[2]

[1] InfoLab, Agricultural University of Athens, Athens, Greece
marios@aua.gr
[2] InfoLab, Agricultural University of Athens, Athens, Greece
tsili@aua.gr

1 Introduction

From a rural land use perspective, an important development in Europe is that agricultural activities are being combined with other activities such as environmental care, maintaining the landscape, forestry, preserving recreational and tourist areas, etc. As a result, there is a strong need for statistical data on rural populations and particularly on landscapes and land use, which are by their nature spatial in form. The management, the processing and the display of such statistical data is therefore, to a large extend, a spatial process. In this respect, GIS is considered necessary in the production of census maps, for dealing with census logistics, for monitoring census activities, and for data dissemination [2].

With the advent of GIS, a wide range of spatial analysis methods has been developed for carrying out data transformations between different spatial structures. These methods help to present the data in a more meaningful and consistent manner and enable different data sets, based on different geographical units, to be brought together and overlaid. They also facilitate the spatial analysis of statistical data required in the development and/or calculation of more reliable indicators for the determination of the state and quality of the environment, and the ability to measure the effect of the agricultural economy, across regions and countries. Most policy makers concerned with agri-environmental issues at the national level are confronted with fragmented information and it is accordingly difficult to use the information in a way that effectively contributes to policy decision making.

A necessary step in the assessment of agricultural policies and of their impact on the countryside and landscapes is the study of spatial units that constitute the underlying structure of these areas. Most statistical data in the European Union (EU), by means of the Farm Structure Survey (FSS) data, is organized and presented on the basis of NUTS (Nomenclature des Unites Territoriales Statistiques) system, to provide a single, uniform breakdown of a country. Nevertheless, these units are geographical areas that may vary substantially not only in their size and shape, but also over time.

This work presents an interface between the statistical and geographical databases and provides a comparison between them by means of the FSS and CORINE Land Cover (CLC). The geographical database can be used as a means for the spatial disaggregation of FSS data into a more accurate geographical level and it is the first step towards a satisfactory spatial analysis. FSS and CLC commonly describe land cover and land use. Definition of an interface between their nomenclatures is a precondition for this spatial disaggregation. Notice that the comparison requires determining the aggregation level of the classes for which the correspondence has already been set, as well as, validation of the result by comparing the respective surface areas of the related classes. After the reclassification of the above data, common classes are created and presented on a map using an embedded GIS environment. FSS data also require a comparison with other sources of information, as for example topography, climatology of the different types of agricultural land, if someone wants, for example, to evaluate the risks of erosion or of pollution of watercourses by pesticides.

Knowing agricultural areas by type of crop within survey districts is insufficient. It is necessary to localize this information more precisely. This will allow the reallocation of data into suitable areas, such as drainage basins, while limiting the loss of information. Notice that land use is difficult to define by photo-interpretation as well as from a large distance (i.e. >100 m). However, this mode of observation is not preponderant concerning unused land, such as, shrub land, forest, bare land, permanent grassland and water/wetland. It concerns south Mediterranean areas where the land cover is very likely to be shrub land or bare lands. On the ground, the high rate of shrub land must reflect a difficulty for defining clearly the activity on such intermediate biotope. These areas can also be considered as unused because of the climate and the low density of population. As the forest unused areas, they might be used as rough grazing areas. From a general point of view, unused areas are much more located on homogenous land cover types (shrub land, land without tree, permanent grassland without tree, forest, etc.). Unused areas occupy a large part of Greece where the proportion rises to almost 40% of the country.

To test the interface and provide the appropriate links between certain classes of the two databases the region of the island of Crete has been chosen. The statistical data used has been provided by the Basic FSS of 1999/2000 (Census of Agricultural for Livestock Breeding or simply Agricultural Census). However, to achieve compatibility between census and photo-interpretation data a recently developed, improved version of the CLC geographical database has been used. The new geo-statistical database, which takes into account the FSS nomenclature and definitions, provides a much better acquisition period

(Landsat-TM 1998–1999) which is the same as the census reference period (1998–1999).

The structure of the paper is as follows: The next section describes briefly the recently introduced geo-statistical database. It provides the main characteristics of the classification scheme used and it resolves the problems encountered when linking the data sources (i.e., the FSS and the CLC databases). Then, the section dealing with the software tool follows, which provides a sufficient description of the development. It should be noted that the developed tool is quite general; however, for validation purposes a case study has been conducted. This section is followed by the section of Data Analysis in which the results from the comparison of the related nomenclatures are presented. Finally, the last section presents the conclusions and discusses further developments of this work.

2 Material and Methods

2.1 The Hellenic Geo-Statistical Database

In the light of recent developments concerning land use statistics and in order to produce more objective information on this sector an up-to-date methodology is adopted using GIS techniques. The National Statistical Service of Greece (NSSG) is testing a methodology to produce a detailed land cover map for the Hellenic territory. The data sources of the land cover map include aerial ortho-photographs, satellite images as well as agricultural census (FSS), and it is based on the same with the CLC minimum mapping unit. The new geo-statistical database aims to cover the needs of land use/cover statistics as far as the distribution of the Hellenic total area into basic categories of land use is concerned. The new database is properly generalized as reference data and harmonized with the FSS nomenclature, by means of characteristics and definitions. As a result, the distribution of the main land uses in Hellas has been organized into 16 classes. For the year 1990, the CLC1990 database is used and therefore a correspondence between the two nomenclatures has been set. Interesting to note that using the 44 CLC classes one may capture the total land cover diversity, i.e. that linked to urban and natural areas. Nevertheless, our interest is to shed light on the relationship between agriculture and the landscape in rural areas and therefore the pre-mentioned reduction in the number of CLC classes is obvious.

Spatial analysis of the information to be recorded is realized by determining the area of the minimum recorded surface, which is taken according to the proposed nomenclature of 16 classes, the methodology of use/cover definition, the requirements of 1:100.000 scale and the user needs. The method, by which the theme information is drawn up, is a comparative photo-interpretation of new satellite data collected in 1998–1999 in relation to those used for the creation of the Hellenic geo-statistical database. The digital photo-interpretation of the new satellite data is made using image processing software and other data such as those from land recordings. The recording, planning and the use of the data from the field work also define the reliability of the specific photo-interpretation.

The new geographical database for the country's area has numerous advantages, the most important of which are the following:

- It provides a land use/cover map covering all Hellenic territory using 16 classes.
- It takes into account the FSS nomenclature and definitions.
- It enables comparability between different time periods, using the same source of information, namely census or photo-interpretation.
- It enables comparability between the two sources of information, namely census versus photo-interpretation. In the case of Hellenic Republic, the acquisition period of the data is spread over 2 years for both, the LTM 1998–1999 and the FSS 1999/2000 (reference year the 1998–1999 crop year).
- It enables the integration of the chrono-geographical co-ordinates of the satellite images sources of CLC. This will help in the identification of districts for which image interpretation is one year apart (minus or plus) from the census year (1990 or 2000, respectively). In addition, using the intermediate FSS data that correspond closely to the date of the satellite image, it will be possible to mitigate the effect of time.

As it appears, the new geo-statistical database is in principle more accurate than CLC. It can be used to calibrate diversity measurements computed from CLC, although there are some problems because the reference dates may not coincide. The methodology has been tested in the region of Crete (NUTS II). The Crete island is about $8,267,\ 45\,\mathrm{km}^2$, it is located in the most south part of Hellenic Republic and it is divided into four administrative areas (NUTS III).

Differences in Data Models

To describe the methodology adopted in the problem we are studying, one has to take into account the non-matching areal units and the problem Modifiable Area Unit (MAUP) [5]. Note that, the temporal incompatibilities problem and the procedure of matching the data points by non-matching due to collection cycles is not considered here.

Starting with the non-matching areal unit problem, as this appears in the pilot case, a new object, called interoperable geo-object is introduced. This object includes all the required procedures in order to solve the following two problems.

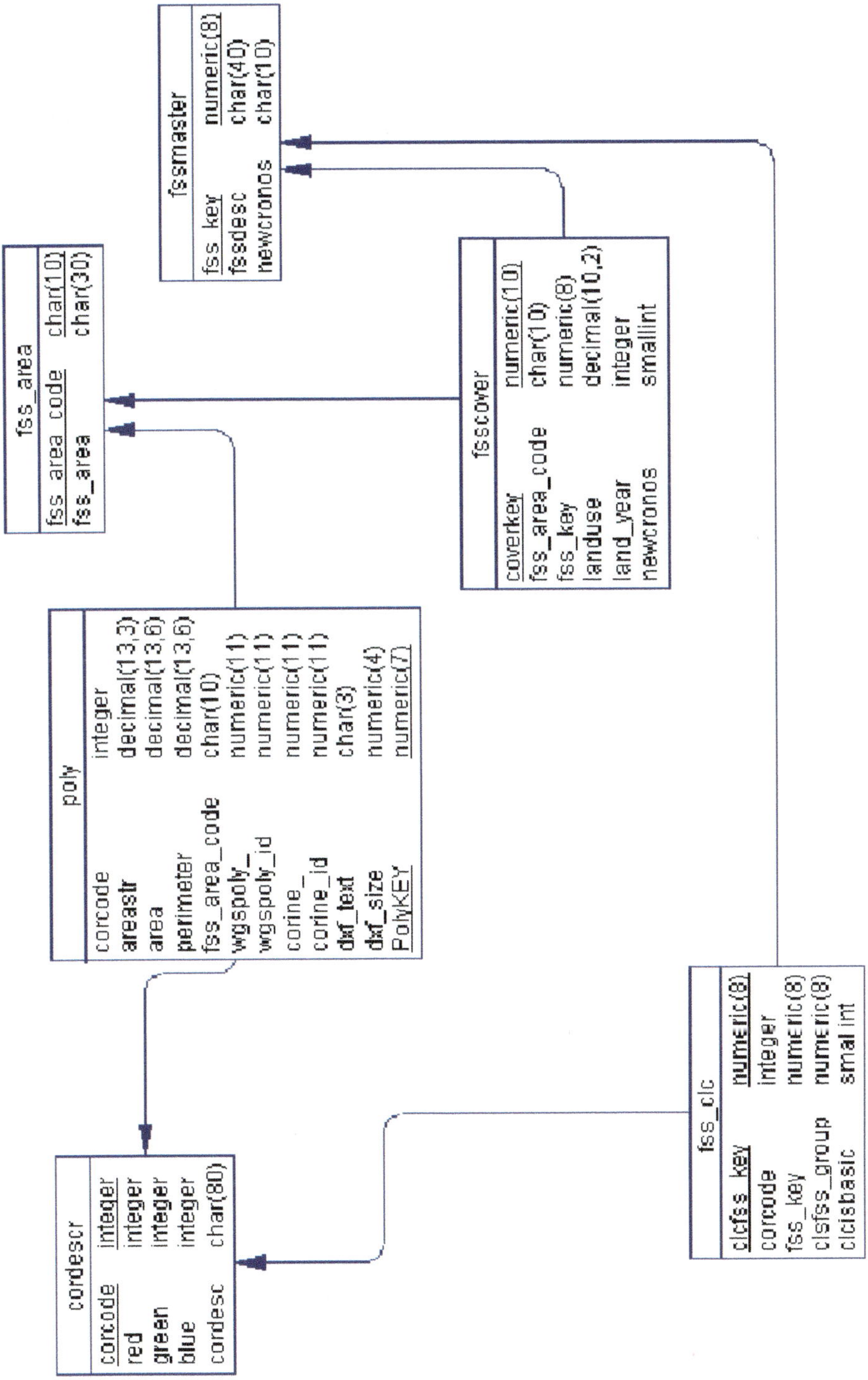

Fig. 1. The conceptual model

- The different boundaries definition of the administrative units that have been used during the collection of the FSS data (1991, 2000) in the pilot regions (NUTS II and NUTS III).
- The geodetic datum used in order to represent jointly the statistical and the ancillary geographical data on a map.

The first problem has been solved with the appropriate transformation between different spatial structures. This transformation determines the process of the aggregation and the disaggregation within nested, nonnested and neighboring polygons. To overlay the data the conceptual model of Fig. 1 has been designed. This model contains and maintains all the polygons and the related geometric data (lines, nodes etc), representing the areal units. To link the descriptive with the spatial information, the data of the geographical area has been divided into smaller parts in order to determine the field that identifies the specific entity (PolyKey), which has been used as a reference key to the GIS. Also, a set of spatial queries has been developed to carry out the above transformation. The second problem has been solved through the development of a common geodetic datum, which represents jointly the statistical and the ancillary geographical data on a map. Finally, an automated procedure has been developed to convert the data from the original to the target geodetic datum.

The MAUP problem has been faced by increasing the spatial detail, using ancillary geographical data [4] such as contour lines, lines representing rivers, or polygons representing lakes etc. This allows the synthesis of geographical data along with the statistical data. Further, it allows the combination of different scenarios to be considered in order to simulate the plotting of the statistical data on a map. For validation and/or prediction purposes, the results have been compared visually with other spatial quantitative information or sampling data presented on thematic maps.

To automate both the transformation between different definitions of administrative units and to achieve the connection between a file containing quantitative data (usually statistical) along with GIS data (ancillary and statistical), two-object classes have been developed namely, the class for data manipulation and the class for GIS manipulation.

3 Software Development – A Case Study

As it has been pointed out, the linkage of the two nomenclatures, by means of the structure survey and the geographical databases, require the development of a software tool able to display maps and descriptive data in a tabular form. This has been achieved by linking the geographical information with the multi-dimensional tabular information of FSS. Thus, the user becomes part of the GIS without the necessity of having specific skills and intimate knowledge of the data used.

The application consists of four items, namely, the relational database, the class of objects for data manipulation, the class of objects for GIS manipulation and the main body of the application software containing the above items along with the functions required by the end user.

To begin with, a step-by-step analysis of the software design is required. The appropriate design steps are described below:

1. The ancillary geographical features, such as contour lines, roads, cities, lakes and rivers are added on the geographical layer of the area of interest. This will help to localize the geographical data.
2. From the FSS database only themes associated with agricultural products have been selected. Note that the use of the geo-object offers the capability to work at different levels of administrative units. However, in the pilot case, the FSS data have been selected at prefecture level (NUTS III), in thousands of hectares, as they are reported in the 2000 census.
3. We develop the entity relationship model as well as the relational database of the software tool, based on the data provided by the FSS database and geo-database.
4. The geographical data have been stored in some database tables of the software tool, using some especially developed functions. Further, the OLEServer method of the QuantitativeInput object has been used with the appropriate DLLs, which have been provided by the FSS, in order to transfer the FSS data into the database.
5. We define the appropriate functions and queries, and we developed object classes in order to achieve uniformity at both the user and the developer levels.
6. We developed an application in which the RDBMS, the GIS and the pre-mentioned object class have been used. The basic capabilities offered by this application are the following:
   - Compose (aggregate) a new FSS theme by selecting one or more classes, and vice versa.
   - Decompose (disaggregate) an existing FSS theme to one or more classes, and vice versa.
   - Correspond (relate) the new FSS themes to classes.
   - Classify (sort) the results by date, county (region), or by class
   - Observe the results plotted on a map and classify these by some geographical characteristics (e.g. allocation of the selected growth by elevation).

4 The Relational Data Base

The GIS tool used for the geo-database construction is the ESRI ArcInfo software. This tool stores a set of tables in DBF format, containing both the spatial and descriptive information about map's features, which are logically organized into themes of information. Each theme consists of topologically linked polygons along with the associated descriptive data. Generally,

X-Base formats, cannot easily aggregate, desegregate, isolate, and combine geographical data with other sources. Furthermore, due to severe limitations associated with the temporal component of data in the GIS raster databases, a comparison between geographical data obtained in the past is very difficult to be achieved in practice [3].

To support the exchange of heterogeneous data into an integrated database environment a conceptual model is required [6]. In the design of such a model one has to take into consideration loading and refreshing the descriptive geographical data for each attribute of the GIS, at any time it will be required, and then linking them with the information derived from other sources, such as the FSS data. The conceptual model of the proposed database is described in Fig. 1.

5 The Classes of Objects for Data and Geo-Database Manipulation

The class of objects for data manipulation is based on a PowerBuilder object called DataWindow [1]. This class provides a simple way of retrieving, displaying and updating data from a specified data source. Although the data source is usually a database, it can also be a text file or other data structure. The class of DataWindow object, named PBDWO, inherits the basic functionality and encapsulates the ability to dynamically, at run time, bind and combine data from different sources.

The geo-database contains only agricultural classes, which can be mapped on one or more regions. To compare them with other ancillary geographical data, for example road, lakes, contour lines, etc., and to process the geographical data, the development of a class of objects that inherit their properties, methods and functions is required. This class of objects encapsulates more functions and customized events to finally communicate with the database, and vice versa. This class is called interoperable geo-object. For the development of this class the ESRI MapObjects is used.

6 The Application Software

As it has been pointed out, this application is computer-based software able to display maps and descriptive data in a tabular form. This has been achieved using geographical information from CLC database linked with tabular information of the multi-dimensional tables of the FSS. The user becomes part of the GIS without the necessity of specific skills and intimate knowledge of the data used.

To test the application software a preliminary study, using the 1991 Basic FSS data of the island of Crete has been prepared. The island of Crete is a region (NUTS II level) and consists of four districts (NUTS III level); Chania,

Rethimno, Iraklio and Lasithi. The geographical presentation is based on the 16 class geo-statistical database (1991 and 2000) and is constructed using the Hellenic Geodetic Reference System 1987 (HGRS 87). Any additional geo-data used such as roads, lakes, contour lines, etc. are constructed using World Geodetic System 1984 (WGS 84). To solve the problem of geodetic datum transformation without making changes in the application source code a map layer object is added. This object has a property to specify the path of the ASCII file, which contains the appropriate transformation parameters. Furthermore, the basic geographic layer is constructed using detailed geographical data, such as coastlines, contour lines, roads, airports etc.

The main application window includes the standard GUI controls (menu and buttons) as well as the PB-DWO and the interoperable geo-object. The PB-DWO contains the rows of the entity cordesc matching the selected area. The interoperable geo-object displays the corresponding polygons of the above entity. Complementary, details of the method followed may be found in Sambrakos et al. [7]. As it may be seen it becomes an easy task for the user to incorporate the geo-statistical and/or the FSS data of any year (and thus for the 2000 year) into the application.

7 Data Analysis

Although the new geo-statistical nomenclature has been harmonized with the FSS nomenclature, there are still some problems related to the two different methodologies. The analysis of the above problems has been carried out by a comparison between the respective areas of the related classes. The available data from the 2000 FSS is based at the Municipality/Commune level (NUTS IV), whereas the data drawn from the new geostatistical nomenclature is at the district level (NUTS III). The data of two databases have been compared in a pilot study of four Hellenic regions at a district (NUTS III) and prefecture (NUTS II) level. The comparison shows large difference between in the agricultural areas. Generally, the examined agricultural areas in the geo-statistical nomenclature are greater than the corresponding agricultural areas in the 2000 FSS. The differences are because of the difficulties in correlating the pastures areas between the two databases, whereas the differences in the arable areas and the areas under permanent crops are related to the different methodologies.

The results found so far are presented in Table 1. Table 1(a) presents the differences (%) in arable areas, areas under permanent crops, and cultivated areas, as they were recorded in the districts (NUTS III) of the examined regions, between the two nomenclatures. Positive sign is in favor of the geo-statistical nomenclature, whereas negative sign is in favor of the FSS nomenclature. Note that the actual differences in the above classes are not as high as they are in the remaining classes, namely agricultural areas (Table 1(b)), pastures and meadows (Table 1(c)), and heterogeneous areas (Table 1(d)). To facilitate the comparison for the last cases the actual values are presented.

Table 1. Results showing the differences between classes, as they have been recorded in the 2000 FSS and the geo-statistical databases

Table 1(a)	districts	(%) differences (2000 FSS – GeoStat)		
(NUTS II)	(NUTS III)	arable areas	permanent crops	cultivated areas
crete	IRAKLIO	−71	4	−4
	LASITHI	54	47	48
	RETHIMNO	−91	−7	−24
	CHANIA	−72	4	−4
total		−66	6	−3
Table 1(b)	Districts	**agricultural areas (ha)**		
(NUTS II)	(NUTS III)	2000 FSS	GeoStat	Differences
crete	IRAKLIO	221,982	139,733	82,249
	LASITHI	127,252	37,864	89,388
	RETHIMNO	115,842	101,182	14,660
	CHANIA	116,472	109,191	7,281
total		581,548	387,970	193,578
Table 1(c)	Districts	**pastures and meadows (ha)**		
(NUTS II)	(NUTS III)	2000 FSS	GeoStat	Difference
crete	IRAKLIO	36,412	69,070	32,658
	LASITHI	16,817	61,631	44,814
	RETHIMNO	62,470	53,241	−9,229
	CHANIA	63,410	40,167	−23,243
total		179,109	224,109	45,000
Table 1(d)	Districts	**heterogeneous areas (ha)**		
(NUTS II)	(NUTS III)	2000 FSS	GeoStat	Differences
crete	IRAKLIO	143	54,339	54,196
	LASITHI	12	34,433	34,422
	RETHIMNO	159	33,372	33,213
	CHANIA	14	32,420	32,406
total		328	154,564	154,237

It has been observed that the above differences in the regions (NUTS II) are generally smaller from the corresponding inter-regional ones (district level; NUTS III). This is due to the fact that the mapping unit of 25 ha in the new CLC is not able to identify parcels of smaller size. This is the case of Greece, in which the average holding size is around 4,5 ha and the average parcel size is around 0,7 ha. An additional reason is that in FSS all the holdings are recorded at the place of residence of the holder (natural person) or headquarter (legal person) of the holding.

8 Conclusions

The work presented so far is a pilot study merging with the use of a software tool the statistical data, available at the administrative level, with the geo-referenced land cover in order to identify and explain the most significant differences encountered between the aggregates of agricultural land cover classes. This has been achieved thanks to the creation of a new geo-statistical database, which is based on both, the FSS and the CLC nomenclature.

The above geo-statistical database seems to provide a good mapping base for Greece, which could be improved further by using suitable satellite images that are able to produce scaled maps of at least 1:50,000. Note that the imposed minimum mapping unit of 25 ha results in an overall underestimation of the diversity of landscapes something, which is particularly important in the case of Greece for which the average size of the holdings is 4,5 ha. Additional sources may be used providing detailed complementary information, such as aerial ortho-photographs, the cadastral map of Greece, IACS (Integrated Administrative Control System), MARS (Monitor Agriculture with Remote Sensing), NATURA2000 database, or other ongoing analysis of the European landscape.

The methodology of using the interoperable geo-object in conjunction with RDBMS settings and the OOP logic means that many of the objects can be used in similar GIS applications with a little effort of maintenance. The application developed is an easy-to-use tool, ideal for comparison of descriptive census results and interpreted geo-data, as well as, to conclude about the correctness of these data. If the expert combines the ability of simultaneous comparison and appearance of results of different years, the conclusions will be more reasonable.

Future research is three fold. Firstly, it is to continue improving the idea of interoperable geo-object by adding methods and properties for uncertainty manipulation and to investigate requirements of GIS in a fuzzy object data model. Our final objective is to provide the geo-object with the ability to generate and visualize transitions from one state to another, using the rules of an expert spatiotemporal system. Related work on this aspect is given in [8]. Secondly, this study may be considered as a first step in the direction of presenting geo-reference statistical and/or agricultural and environmental data. As soon as this initiation will be completed, it will become possible to redistribute quantitative data other than land use from the FSS by defining some distribution rules using co-variables. Finally, this research will facilitate the spatial analysis of statistical data required in the development and/or calculation of more reliable indicators.

References

1. Candak R., Chandak P. (1999) Advanced in PowerBuilder 7 Techniques. Wiley, New York, pp. 202–253.
2. Deichmann U. (1997) Geographical information systems in the census process Technology options, costs and benefits. Workshop on Strategies for the 2000 Round of Population and Housing Censuses in the ESCWA Region, Cairo, December 6–10.
3. Dragicevic V., Marceau D. (2000) An application of fuzzy logic reasoning for GIS temporal modeling of dynamic processes. Fuzzy sets and Systems Vol 113: 69–80, Elsevier.
4. Flowerdew, R., Green M. (2001) Areal Interpolation and Types of Data. In Stewart Fortheringham and Peter Rogerson (eds) Spatial Analysis and GIS. London, Bristol: Taylor & Francis Ltd.
5. Openshaw S. (1984) The Modifiable Areal Unit Problem In Norwich: Geo Books, Norwich.
6. Parent C., Spaccapietra S., Zimanyi E. (1999) Spatio-temporal conceptual models: Data structures + space + time. Proceedings of the 7th ACM Symposium on Advances in Geographic Information Systems, GIS'99, Kansas City, USA.
7. Sambrakos M., Fillis I., Tsiligiridis T. (2001) Integration of Spatial Descriptive Statistical Data and Geographic Information. Proceedings of 8th Pannellenic Conference on Informatics, paper 91C, Nicosia, Cyprus, November 2001.
8. Sambrakos M., Yialouris C., Tsiligiridis T. (2002) Fuzzy Interoperable Geographical Object (FIGO). An approach of enhancing spatial objects with fuzzy behavior. Proceedings of the 1st Conference of HAICTA (Hellenic Association of Information and Communication Technology in Agriculture, Food and Environment), Session 3B, pp 209–219, Athens University of Agriculture, Athens, Greece, June 6–7, 2002.
9. Sambrakos M., Tsiligiridis T. (2003) A Comparative Landscape Pattern Analysis Using Remotely-Sensed and Statistical Data to Evaluate Regional Diversity. Proceedings of 1st International Congress of ITAFE 03 (Information Technology in Agriculture, Food and Environment), O-GIS-33-0, Edge University, Izmir, Turkey, October 7–10, 2003.
10. Yuan Y., Smith R., Limp W. (1997) Remodeling Census Population with Spatial Information from Landsat TM Imagery. Computers, Environment and Urban Systems, 21:245–258.

Epidemiological Information Systems

V. Gómez-Rubio[1], J. Ferrándiz-Ferragud[2], and A. López-Quílez[3]

[1] Dpto. Matemáticas, Universidad de Castilla-La Mancha, Albacete, Spain
Virgilio.Gomez@uclm.es
[2] Dpto. Estadística e Investigación Operativa, Universitat de València, València, Spain
Juan.Ferrandiz@uv.es
[3] Dpto. Estadística e Investigación Operativa, Universitat de València, València, Spain
Antonio.Lopez@uv.es

1 Introduction

1.1 GIS and Public Health

Public Health Authorities can take advantage of GIS in order to perform studies in disease surveillance tasks, to know how diseases spread and to locate outbreaks. Facts involved in these studies come from a wide range of sources: hospital registers, clinicians, environmental organisations, etc., so it is important to collect and store all them for it being easier to access and analyse.

The increasing interest in GIS in Public Health has also been reflected in the literature. In last years, a number of books have appeared devoted to GIS and Public Health affairs [4, 9, 15], statistical methods for Spatial Epidemiology [19, 23, 24], and several journals, such as *Statistics in Medicine* [11, 30] and *Journal of Royal Statistical Society* [32] have devoted special issues to related subjects.

1.2 A GIS for Spatial Epidemiology

Since the range of applications of GIS in Public Health is nearly unlimited (like in many other fields), we will focus on Spatial Epidemiology, which refers to different topics about the spatial spread of diseases: disease mapping, detection of clusters of disease, ecological analysis, etiology, etc.

In this paper we describe how a Geographic Information System for Spatial Epidemiology can be developed and we briefly discuss the main points to which attention should be paid.

In Sect. 2 we describe the main issues concerning data. Section 3 covers statistical methodology. Section 4 comprehends how the whole GIS can be developed, integrating and analysing data. What we have developed is explained in Sect. 5.

2 Managing Data

Data needed are determined by the conclusions we want to draw from the studies. Usually, the main concern is to explore the spatial distribution of a group of diseases (mortality and/or morbidity), which is accomplished by means of *Disease Mapping* (see Sect. 3.2). The second step is often the detection of those regions where there exists a higher risk of suffering from these diseases, known as *Risk Assessment* (as explained in Sect. 3.3).

When a high risk has been detected an explanation is usually required. Sometimes it is possible to look for relationships between risk and a number of covariates. This is done via *Ecological Analysis*, as described in Sect. 3.4.

Studies are always restricted to a period of time and to a particular area. Data available are aggregated on the basis of units used to measure space and time. For this, year is often used, while there is no clear preference for the spatial units, since it usually depends on administrative boundaries.

The level of aggregation is restricted due to confidentiality issues. Data available in the studies are usually in a form that prevents from identifying single individuals. This means that short periods of time or quite small areas can't be used.

Data quality uses to be quite good for mortality, but it doesn't happen to morbidity, excepting for a few set of 'important' diseases for which special registers are drawn up (like, for example, children malformations, AIDS or cancer).

Measuring exposure to a risk factor is always difficult and it is often impossible to take exact measurements for every person at risk. Residence is often used as a proxy to the place of exposition, although it can be misleading since people is also expected to spend quite time at work, for example.

Nevertheless, government agencies and other data providers usually link their information to the appropriate administrative areas, be it quarters, electoral districts, etc. When this doesn't happen it is not difficult to refer the actual location to the standard administrative areas [27].

Finally, it is important to update data on a regular basis, so that recent problems can be investigated.

2.1 Population Data

Population at risk is a key issue, since it helps to measure the incidence or prevalence of a disease. When calculating risk rates, a ratio between the number of affected people and population at risk is calculated (see Sect. 3.1).

Inaccuracies in the denominator can lead to wrong estimations, especially in low populated areas.

Measurement of population at risk is even harder than morbidity or mortality since it may not be clear how to define *at risk*. The most common approach is to consider residence as the main place where people stay, hence where they are exposed. But it can be inaccurate since work is an important place where people can be exposed. This is especially important when the disease under study may be related to the kind of employment.

Migration is another source of bias. Information about residence is collected on a regular basis every ten years by census offices. If population of any year in between is needed an estimation is carried out [29]. Taking migration into account is difficult since there are no clear clues about how to approach it because the high number of economic, social and political factors that may be influencing.

2.2 Health Data

Health data are collected by a number of health institutions, such as hospitals, and gathered by the main Local Health Authority. Usually, mortality is recorded with a high level of accuracy, while morbidity is more difficult to obtain.

For those diseases of high interest, such as AIDS or Cancer, separated records are maintained by a specialised institution (for example, Cancer Registries) and provided information is more accurate.

2.3 Geographic Data

The previously described information is always referred to an administrative level and, if not, it is not difficult to make a conversion so they are. Since administrative levels are usually structured into layers (from smaller to larger areas), knowing how to aggregate (or disaggregate) data can help to conduct studies at different levels.

Administrative boundaries can be used to plot maps that summarise results from analysis. Cloropleth maps are the most common representation, in which the variable under study is categorised and every administrative region is filled with a colour according to these categories.

2.4 Environmental Data

Sometimes, disease outbreaks are related to environmental conditions. Several issues may increase the risk of suffering from certain diseases. For example, exposure to pollutants of a nearby petrochemical complex may increase the incidence of leukaemia. For this case, distance to the putative source can be used as an approximation to the level of exposure [27].

This kind of data usually comes from a wide range of sources and variables of interest depend on the kind of study to be carried out. Some of them may be available from government agencies but others must be collected ad hoc for the studies.

Maps can be created to represent the spatial distribution of these variables (see Fig. 1). When a continuous representation is required, geostatistical methods [8] can be used to provide estimations at those points where a sample hasn't been taken.

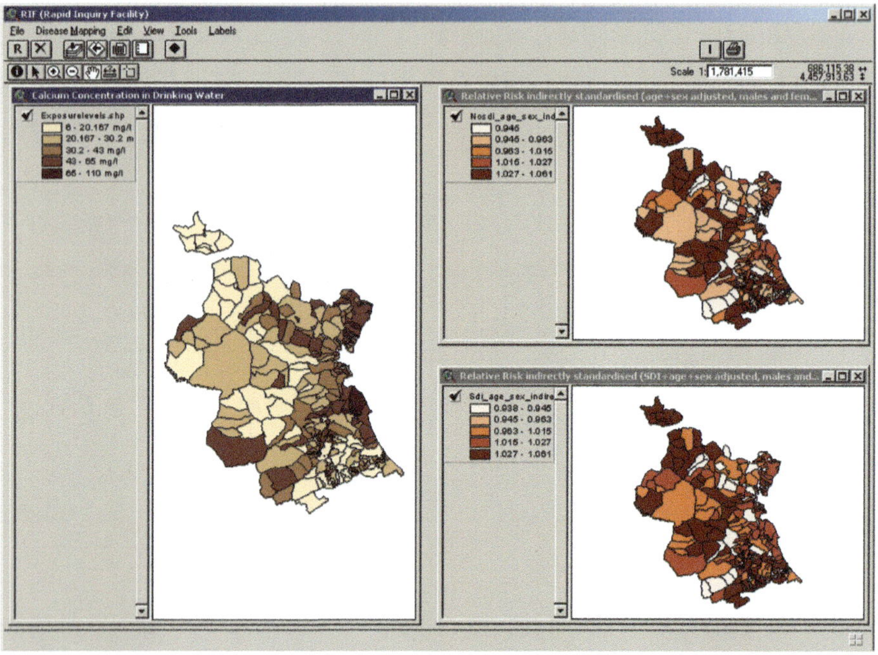

Fig. 1. Maps produced by the RIF (described in Sect. 5) in an investigation of the relationship between hardness of drinking water and cerebrovascular mortality in the province of Valencia (Spain). On the *left*, we have the spatial distribution of calcium in drinking water, while on the *right*, SMRs (*top*) and SMRs standardised also by calcium levels (*bottom*) are shown

2.5 Socio-Economic Data

Social and economical factors may also influence results. It has been proved that way of life, diet, deprivation, etc. may have a strong influence over disease risk. A deprivation index is created taking into account different variables and it is often used in standardisation to filter socio-economic inequalities [12].

2.6 The Right Data for the Right Analysis

Once all these data have been collected, the next step is to make them available from a common source. This compilation implies a debugging of the data in order to detect and correct (or remove) wrong entries. This point is crucial since data quality is the key to what kind of studies can be carried. Imputation methods must be used to fill in missing data.

The same database should be used to store all the information available in order to integrate different types of data and a link among them must be established. This is done by means of the spatial location of the data, so it is necessary to unify how it is specified: coordinates, municipalities, counties, etc. This is a challenging task since there can be a lot of different spatial units.

When data are measured at individual locations (i.e., different measures at single, maybe different, points) a mechanism to associate these points to administrative regions may be needed [27]. Point data can be analysed, but following a different approach from the one employed for area based data.

3 Statistical Analysis

3.1 General Analysis

Although standard and simpler methods can be used to carry out a number of analysis, we should take advantage of the spatial nature of our data and the special characteristics of the problem we are facing [8, 19].

As a rule, cases (mortality and/or morbidity) and population are stratified according to sex, age group and, if available, a measure of deprivation or poverty. These factors have been proved to be important in the analysis, and this approach helps to reduce bias in the estimations and to remove the effect of this factors [22].

We will also find different measures for administrative regions and periods of time, so that we have variables O_{ijk} and P_{ijk}, for the observed number of cases and population, respectively, in region i, age-sex-deprivation strata j and period of time k. When working with fully spatial models, the third component will be missed and data will be aggregated based on the third component.

When calculating incidence rates, a quotient between affected people (numerator) and population at risk (denominator) is made [22]. When a comparison between different studies at different locations is required, another region is used as a reference to calculate standard rates for each age-sex-deprivation stratum.

If we call r_{jk} the reference (or comparison) rate for stratum j and period k and supposing no region – age-sex-deprivation interaction [34], we can get an estimation of the expected number of cases in region i and period of time k by $E_{ijk} = r_{jk} \cdot P_{ijk}$. This process is called *standardisation* [22].

Most models suppose O_{ijk} drawn from a Poisson distribution whose mean is $\theta_{ik}E_{ijk}$, where θ_{ik} is called the *Relative Risk* [34]. Its Maximum Likelihood Estimator is $\widehat{\theta}_{ik} = \sum_j O_{ijk}/\sum_j E_{ijk}$, called the *Standardised Mortality Ratio* (S.M.R.). Usually, a confidence interval is calculated to test significant departure from value 1, which marks the standard risk.

Another layer can be added to the model in order to explain how θ_{ik} is distributed (see Sect. 3.5). Usually, a Generalised Linear Model is constructed [26], although Bayesian Hierarchical Models with spatial structure have been used to smooth relative risks (as shown in Sect. 3.5), so that neighbouring regions are also taken into account in the local estimations.

3.2 Disease Mapping

Disease Mapping is used to display the spatial distribution of a disease. Usually, a statistic associated to the disease is calculated and later split into different categories. Then, areas belonging to the same category are filled with the same colour (see Fig. 1).

Although these maps can be really helpful as an early summary, they can be misleading, since higher areas will attract more attention and low populated areas tend to show more extreme values [6]. Some authors have also addressed other problems related to the influence of scale, map projection and colours used [9].

Common statistics to be mapped include p-values [5], S.M.R.s [34] and residuals from a fitted regression model [8] (usually used to see whether any spatial structure remains unexplained by covariates).

3.3 Clusters of Disease

By taking a look at a disease map, groups of areas with higher risk (*clusters*) can be detected. Due to the problems mentioned in the previous subsection, this method can be misleading and not accurate.

Since the detection of clusters of disease is one of the priorities for epidemiologists, a number of methods have been developed for this purpose, and a few reviews have been made by authors in the last years [19, 25, 33].

In the investigation of clusters of disease we can mainly distinguish two types: search for clusters in the study area [33] or investigating a known putative pollution source [10]. Clearly, statistical assumptions are quite different depending on which one we are working on.

3.4 Ecological Regression Analysis

The relationship between disease and risk can be investigated through *Ecological Regression*. Generalised Linear Models [26] are often used for this purpose, although Generalised Additive Models has also been used [20]. An example of

Ecological Regression using GLMs is the study about the relationship between hardness of drinking water and cerebrovascular mortality [14].

When performing an Ecological Regression, it is important to pay attention to how risk exposure has been taken. If different levels of aggregation are used it may happen that some measurements have been made at a broad level, i.e., the same risk is associated to a wide range of population and a bias may be introduced in the analysis. This problem is know as the *Ecological Fallacy* [28].

3.5 Bayesian Hierarchical Models

The Bayesian paradigm has been successfully applied to all fields of Spatial Epidemiology. It is based on Bayes' Theorem, so that we calculate the posterior distribution (after observing the data) as the product of likelihood and priors of the random variables. When the posterior distributions can't be worked out, as it happens most times, MCMC techniques [16] are employed to simulate them.

Probably, the first Bayesian Hierarchical Model applied to Spatial Epidemiology was the one proposed by Clayton and Kaldor [6], which proposes a prior Gamma for all the relative risks and produces a globally smoothed estimate of the relative risk $\widehat{\theta}_i = (O_i + \nu)/(E_i + \alpha)$, which is a compromise between the S.M.R. of region i and the prior mean (ν/α), reducing extreme values.

Other models, such as the one proposed by Besag et al. [2], also produce smoothed estimates of relative risks. The logarithm of the relative risk is expressed as the sum of the effect of neighbouring areas plus the effect of the local area, which can be a linear function of covariates [13]. Then, the estimations of the relative risks obtained have been smoothed by taking into account the effect of neighbours.

Smoothed estimates of relative risks obtained in these models can be used to produce cloropleth maps. Comparing these maps to those made from SMRs will show how the effect of extreme values in low population areas is reduced.

4 Integrating Data and Statistics

4.1 Previous Consideration

First of all, it is necessary to know current and future needs before designing the whole system. Perhaps, we only must care about the spatial spread of diseases, without paying attention to possible causes. Or, on the contrary, the main concern is to investigate sources of pollution to see how they affect health. This is important because data and statistical methods will depend on these needs.

For basic models with few data, they could be imported into any statistical software available, while for huge amounts of data, statistical methods will require better integration with the database in order to make analysis possible in a short time (or even possible!).

Time required for the analysis may also be important. Exploratory tools may be used as a first, rapid look into a problem to decide whether a further investigation should be carried out. If it is decided so, more time consuming methods (such as, for example, Bayesian methods computed via MCMC) may be required for a more accurate analysis.

4.2 Managing Data

Since data are collected from many different sources, it is crucial to integrate them into a single database. Before doing this, data quality must be assured and it should be checked for inadequacies, wrong and missing values, etc. Data can be linked by referring to their spatial location, as explained in Sect. 2.

Health data can be stored as single events, but usually a minimal aggregation is defined for space and time, and data will be aggregated on the fly when performing a study. Depending on the amount of data available it may be useful to create separated tables of aggregated data to speed up future investigations.

It is important to provide a way to move up and down the different administrative layers in order to be able to carry out studies at different levels. This can be done by providing a conversion table from one level into another.

4.3 Computing Statistics

Basic statistics can be computed with any statistical software available. Although some GIS software are incorporating statistical methods for spatial statistics, they are mostly focused on geostatistics and it is difficult to find methods used in Spatial Epidemiology.

Accessing data directly to the database can solve this problem but, as commented before, when the amount of data is big, it can be very slow and it should be better to implement statistical methods inside the database.

Unfortunately, as stated in [9], there is a lack of software in the field of Spatial Epidemiology, and what can be found so far, are isolated programs to compute a few methods.

4.4 Interfacing Data and Statistics

Although GIS, databases and statistical software can be used separately when performing studies, it can be more helpful (and less time consuming) to develop an unified tool.

An interface can be created to define the investigation to be carried out, query the database and perform the statistical computations. Most GIS provide a programming language which enables the development of internal routines. Sometimes, it is possible to link the GIS to statistical software and the database, so that all results are returned to the GIS and, for example, cloropleth maps can be developed.

Statistical outcomes can also be displayed in reports, and results are often grouped by region and age-sex-deprivation stratum. This is useful when a comparison among them is required, although a Comparison Test could be used for this purpose [12].

5 Software Developed

5.1 Rapid Inquiry Facility

The Rapid Inquiry Facility (RIF) was initially developed at the Small Area Health Statistics Unit [1], but it was rewritten within the framework of EUROHEIS Project [7], funded by the European Commission. It was intended to be a Health Information System for Disease Mapping and Risk Assessment around putative pollution sources.

It is based on ArcView 3.2, Oracle Database 8i, and Oracle Forms and Reports. A graphical user interface developed in Avenue (ArcView's internal programming language) allows the selection of study and comparison regions together with the period of time and diseases to investigate. Two types of studies can be done: *Disease Mapping* and *Risk Assessment* around putative pollution sources.

The RIF was developed with a standard structure (see Fig. 2) in order to allow all the partners of the project to customise it at their Local Health

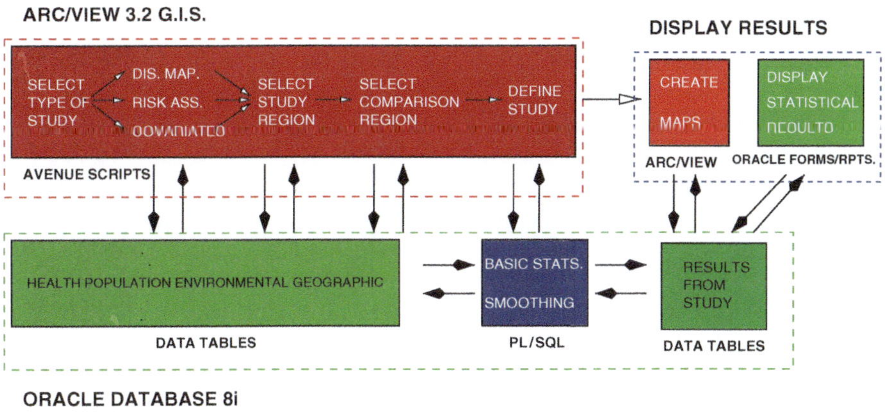

Fig. 2. RIF structure

Authorities with their own data. As the Spanish Partner, we did it at the Conselleria de Sanitat (Comunidad Valenciana, Spain).

As numerator tables, we have mortality, and hospital admissions (removing repeated registers) are used as a proxy to morbidity. Population, as provided by Census register, is used as denominator. All these data are available from 1989 to 1998 at the level of municipality, due to confidentiality.

A deprivation index [12] was also developed and incorporated into the system to be used, together with sex and age group, in standardisation.

While mortality and morbidity are stored as single cases in the database (and later aggregated in the studies) population is stored as habitants per age-sex-deprivation stratum in each municipality.

Administrative boundaries at three levels (municipality, province and autonomous community) are available in the system, and any of these levels can be used when carrying out a study. They are stored as shapefiles, which are used by ArcView to create maps. The code of the municipality is used to link different data available.

For Disease Mapping, basic statistics (expected cases, observed cases, relative risk and its 95% confidence interval) are calculated for each region under study and Poisson-Gamma Model [6] is used to provide smoothed estimations of relative risks.

For Risk Assessment, one or several points regarding putative pollution sources (nuclear plants, waste incinerators, tile industry, etc.) are selected, and regions are grouped according to their distance to these points and the same basic statistics as before are calculated.

Beside maps based on the results, a report with all the statistics is created by Oracle Forms and Reports. Results are grouped by administrative region and 6 groups depending on sex (males, females and both sexes) and type of standardisation (whether deprivation index is taken into account). This way of presenting data is useful to compare risk between different sex and deprivation levels. All studies are stored in the database so they can be accessed later.

5.2 Enhancing the RIF

Although the RIF provides a rapid look into the data, we missed a few capabilities in the system. For example, it doesn't perform any test to compare results among different sex-deprivation strata, which is quite important. Furthermore, covariates can't be used in the studies and there is no possibility of exporting results to be analysed with an external statistical software.

The use of covariates in the study was implemented inside the RIF using the existing structure. Covariates are treated as sex, age group and deprivation index in standardisation [12]. Each covariate is split into groups defined by the user and rates are calculated with and without standardising by covariate groups. If the covariate really has any relationship with the disease, we expect to have different results.

For further statistical analysis we decided to use the R Statistical Environment [21] because it provides a wide range of options for spatial analysis and it is freely available. We developed a couple of packages, *RArcInfo* [17] and *DCluster* [18] to be used for Disease Mapping and detection of clusters of disease, respectively.

In order to use Bayesian Hierarchical Models, WinBUGS [31] is a suitable software. Although MCMC calculations must be used and analysed with care, a few models can be predefined so that they are automatically run after plugging-in results from the RIF. This can be done within R, since a link to WinBUGS has been developed and convergence of Markov Chains can be checked in R by using package CODA [3].

Acknowledgements

This work has been partly funded by Conselleria de Sanitat, Comunidad Valenciana (Spain) and EUROHEIS Project, code SI2.329122 (2001CVG2-604), funded by European Commission.

References

1. P. Aylin, R. Maheswaran, J. Wakefield, S. Cockings, L. Jarup, R. Arnold, G. Wheeler, and P. Elliot (1999) A national facility for small area disease mapping and rapid initial assessment of apparent disease clusters around a point source: The U.K. Small Area Health Statistics Unit. *Journal of Public Health Medicine*, 21(3):289–298.
2. J. Besag, J. York, and A. Mollie (1991) Bayesian image restoration, with two applications in spatial statistics. Annals of the Institute of Statistical Mathematics, 43:1–59.
3. N. G. Best, M. K. Cowles, and S. K. Vines (1995) *CODA Convergence Diagnosis and Output Analysis software for Gibbs Sampler output: Version 0.3.*
4. D. J. Briggs, P. Forer, L. Järup, and R. Stern, editors (2002) *GIS for Emergency Preparedness and Health Risk Reduction.* Kluwer Academic Publishers.
5. M. Choynowski (1959) Map based on probabilities. *Journal of the American Statistical Society*, 54(286):385–388.
6. D. Clayton and J. Kaldor (1987) Empirical bayes estimates of age-standardized relative risks for use in disease mapping. *Biometrics*, 43:671–681.
7. S. Cockings and L. Järup (2002) A European Project and Environment Information System for exposure and disease mapping and risk assessment (EUROHEIS). In D. J. Briggs, P. Forer, L. Järup, and R. Stern, editors, *GIS for Emergency Preparedness and Health Risk Reduction*, volume 11 of *NATO Sciences Series*, chapter 11, pages 201–226. Kluwer Academic Publichers.
8. N. A. C. Cressie (1993) *Statistics for Spatial Data.* John Wiley & Sons, Inc.
9. E. K. Cromley and S. L. McLafferty (2002) *GIS and Public Health.* The Guilford Press.
10. P. Diggle and P. Elliot (1995) Disease risk near point sources: Statistical issues for analyses using individual or spatially aggregated data. *Journal of Epidemiology and Community Health*, 49: S20–S27. (Suppl. 2).
11. K. H. Falter, D. R. Betts, D. B. Rolka, H. R. Rolka, and W. K. Sieber, editors (1999) *Statistics in Medicine*, volume 18 (23). Wiley and Sons, Inc. Special Issue: Symposium on statistical bases for Publich Health decision making: from exploration to modelling.
12. J. Ferrándiz, J. J. Abellán, V. Gómez-Rubio, A. López-Quílez, P. Sanmartín, C. Abellán, M. A. Martínez-Beneito, I. Melchor, H. Vanaclocha, O. Zurriaga,

F. Ballester, J. M. Gil, S. Pérez-Hoyos, and R. Ocaña (2004) Spatial analysis of the relationship between cardiovascular mortality and drinking water hardness. *Environmental Health Perspectives*, 112(9):1037–1044.
13. J. Ferrándiz, J. J. Abellán, A. López, P. Sanmartín, H. Vanaclocha, O. Zurriaga, M. A. Martínez-Beneito, I. Melchor, and J. Calabuig. Geographical distribution of cardiovascular mortality in Comunidad Valenciana (Spain). In D. J. Briggs, P. Forer, L. Järup, and R. Stern, editors (2002) *GIS for Emergency Preparedness and Health Risk Reduction*, volume 11 of *NATO Sciences Series*, chapter 15, pages 267–282. Kluwer Academic Publichers.
14. J. Ferrándiz, A. López, V. Gómez-Rubio, P. Sanmartín, M. A. Martínez-Beneito, I. Melchor, H. Vanaclocha, O. Zuriaga, F. Ballester, J. M. Gil, S. Pérez-Hoyos, and J. J. Abellán (2003) Statistical relationship between hardness of drinking water and cerebrovascular mortality in Valencia: A comparison of spatiotemporal models. *Environmetrics*, 14(5):491–510.
15. A. Gathrell and M. Löytönen, editors (1998) *GIS and Health.* Number 6 in GISDATA. Taylor & Francis.
16. W. R. Gilks, S. Richardson, and D. J. Spiegelhalter, editors (1996) *Markov Chain Monte Carlo in Pratice.* Chapman and Hall.
17. V. Gómez-Rubio and A. López-Quílez (2005) RArc Info: Using GIS data with R. *Computers & Geosciences* 31:1000–1006.
18. V. Gómez-Rubio, J. Ferrándiz-Ferragud, and A. López-Quílez (2005) Detecting clusters of disease with R. *Journal of Geographical Systems* 7(2):189–206.
19. R. Haining (2003) *Spatial Data Analysis. Theory and Practice.* Cambridge University Press.
20. T. Hastie and R. Tibshirani (1986) Generalized additive models. *Statistical Science*, 1:297–318.
21. R. Ihaka and R. Gentleman (1996) R: A language for data analysis and graphics. *Journal of Computational and Graphical Statistics*, 5(3):299–314.
22. M. Jenicek and R. Cléroux (1982) *Epidemiologie. Principes, Techniques, Applications.* Edisem Inc., 2 edition.
23. A. Lawson, A. Biggeri, D. Böhning, E. Lesaffre, and J.-F. Viel, editors (1999) *Disease Mapping and Risk Assessment for Public Health.* John Wiley & Sons Ltd.
24. A. B. Lawson (2001) *Statistical Methods in Spatial Epidemiology.* Wiley.
25. R. J. Marshall (1991) A review of methods for the statistical analysis of spatial patterns of disease. *Journal of the Royal Statistical Society, Series A*, 154(3):421–441.
26. P. McCullagh and J. Nelder (1989) *Generalized Linear Models.* Chapman and Hall.
27. J. Pekkanan, E. Pukkala, M. Vahteristo, and T. Vatiainen (1995) Studying cancer incidence around an oil refinery as an example of a small area study based on map coordinates. *Environmental Research*, 71(2):128–134.
28. S. Piatandosi, D. P. Byar, and S. B. Green (1988) The ecological fallacy. *American Journal of Epidemiology*, 127(5):893–904.
29. P. Rees, P. Norman, and D. Brown (2004) A framework for progressively improving small area population estimates. *Journal Royal Statistical Society A*, 167:5–36.
30. W. K. Sieber, T. A. Green, G. S. Haugh, M. jo Kresnow, E. T. Luman, and H. G. Wilson, editors (2001). *Statistics in Medicine*, volume 20 (9–10). Wiley

and Sons, Inc. Special Issue: Symposium on Emerging Statistical Issues in Public Health for the 21st Century.

31. D. J. Spiegelhalter, A. Thomas, and N. G. Best (1999) *WinBUGS Version 1.2 User Manual.* MRC Biostatistics Unit.
32. J. Wakefield, M. Quinn, and G. Raab, editors (2001) *Journal of the Royal Statistical Society (Series A)*, volume 164 (1). Blackwell.
33. J. C. Wakefield, J. E. Kelsall, and S. E. Morris (2000) Clustering, cluster detection and spatial variation in risk. In P. Elliot, J. C. Wakefield, N. G. Best, and D. J. Briggs, editors, *Spatial Epidemiology. Methods and Applications.*, chapter 8, pages 128–152. Oxford University Press.
34. J. C. Wakefiled, N. G. Best, and L. Waller (2000) Bayesian approaches to disease mapping. In P. Elliot, J. C. Wakefield, N. G. Best, and D. J. Briggs, editors, *Spatial Epidemiology: Methods and Applications*, pages 104–127. Oxford University Press.

Geomatics, Epidemiology and BioStatistics: An Application to Acute Coronary Syndrome

Théophile Niyonsenga[1], Josiane Courteau[2], Charmaine Dean[3], Abbas Hemiari[2], Goze Bénié[4], and Alain Vanasse[2,5]

[1] Robert Stempel School of Public Health, Florida International University (FIU), Miami, FL (USA)
theophile.niyonsenga@fiu.edu
[2] PRIMUS group, Clinical Research Center, CHUS, Sherbrooke (QC), Canada
[3] Statistics and Actuarial Science, Simon-Fraser University, Vancouver, BC, Canada
dean@stat.sfu.ca
[4] Geography and Remote-Sensing Department, Université de Sherbrooke, Sherbrooke, QC Canada
Goze.Bertin.Benie@USherbrooke.ca
[5] Family Medicine Department, Université de Sherbrooke, Sherbrooke (QC), Canada
alain.vanasse@usherbrooke.ca

1 Introduction

The emerging field of Geomatics has found useful application in several research areas. A new phase of its development merges it into the realm of epidemiology and public health to bring insight into the regional disparities of disease incidence and for disease surveillance. New advances in biostatistics include spatial statistics methods which aim to specifically understand and model the spatial variability. Spatial statistics, when combined with geomatics, constitutes an excellent and powerful analysis approach to handle and better understand health issues, specifically in disease related prevention and intervention studies.

The main goal of this paper is to explore the integration of geomatics and spatial statistics with an application to a specific health issue. The outcome of interest is acute coronary syndrome (ACS) incidence in the province of Quebec (Canada) between 1996 and 1998, and hospital readmission at one month post-discharge. It is an established fact that mortality and hospital-acquired infection rates are indicators of the quality of care [24] but more recently some studies are turning their attention to the early hospital readmission rate [2, 26]. Within this context, following specific questions are addressed: Is there spatial heterogeneity and/or spatial aggregation in the ACS incidence and early readmission rates? Is there any geographical trend in the rates? Is there an explanation for the spatial heterogeneity? By ACS, we mean the occurrence of myocardial infarction (MI) or unstable angina. By early readmission, we mean readmission of a patient for coronary heart disease (ACS and angina) 30-days post discharge. Although practice guidelines have been in circulation to standardize the treatment and follow-up of acute myocardial infarction [16, 22], regional variations are currently reported in the literature [20, 21, 25]. The presence of a complex network of factors influencing care quality [2], hospital readmission and the interaction between them have been put forward as the potential explanation of the observed spatial variability in hospital readmission rates. Most of these factors center on the patient while the data available and/or the interests of public policy makers focus on rates of local health units over a given time period. Over these administrative geographical units, interest centers on variables that could explain spatial variability, such as deprivation indices and other area specific characteristics, and help to understand inequalities within health care services and accessibility.

2 Methods

2.1 Population

The studied population consists of all the patients living in the Québec province in Canada, hospitalized for an ACS. The first registered hospitalization during the study period (1996–1998) will be considered as the "index hospitalization". The Québec register "Maintenance et Exploitation des Données pour l'Étude de la Clientèle Hospitalière (MED-ECHO)" made possible the identification of the patients that fit inclusion criterion. This register lists all summary administrative data collected when any patient is treated in an acute care hospital in the Québec province. The validity of this data, concerning MI, was studied and its results published [14, 19, 27, 28]. The inclusion criterion is the inscription of the code 410 (MI) or 411 (unstable angina) of the international classification of disease 9th revision (IDC-9) as the main diagnosis for the hospitalization. In order to increase the study's internal validity, we excluded patients with an error of code of residence and patients that were younger than 25 years old, because they are more likely to have had a MI caused by a different pathophysiological process. For the same reasons, we wanted 'new' cases of ACS, so we excluded patients that were hospitalized with an ACS in the year preceding the index hospitalization.

2.2 Geographical Unit

Spatial data sources used in this study are ESRI [7], DMTI Spatial [5] and the Ministère de la santé et des services sociaux [3, 17]. The geographic coordinate system used in mapping is GCS North American 1983. Spatial data and cartographic representations were managed using ArcGIS [1].

2.3 Variables

The hospitalization for ACS and early hospital readmission are our dependent variables. The former variable is defined as the first occurrence of an ACS hospitalization (main diagnosis IDC-9 codes 410 or 411) in the 3 years' study (index hospitalization), and the latter is defined as a early readmission for a coronary heart disease (main diagnosis IDC-9 codes 410 to 414) in the 30-days following the index hospitalization. The incidence and the readmission rates were calculated by Local Health Unit (LHU). For the readmission rate, we excluded all in-hospital deaths and all deaths encountered within 30-days post discharge. Two deprivation indices were retained as potential explanatory variables for the spatial heterogeneity in health outcomes. Pampalon et al. [18] calculated two sets of deprivation quintiles for approximately 9,000 enumeration areas (EA) using a principal component analysis (PCA). The two principal components of this PCA reflected a material dimension of deprivation (percent of people without a secondary certificate, the ratio employment by population, the average income) and a social dimension of deprivation (the percent of people divorced, separated or widowed, the percent of single-parents, the percent of persons living alone). For each of these PCA components, the EA were ordered by their factor score – from the least to the most deprived – and then the population was fragmented in quintiles (the fifth quintile being the most deprived). For each LHU, the population that belongs to each quintile was calculated. Based on these quintiles, we defined two variables, denoted by material deprivation index (MDI) and social deprivation index (SDI), by the percent of the LHU population that belongs to the fifth quintile.

2.4 Analyses

The analyses were performed in two steps. We first focussed on the geographical distribution of rates (spatial heterogeneity and clusters). To detect clusters and particular hot spots, we used SaTScan [12]. To see the general spatial trend in the rates, we used the Geographically Weighted Regression [8] approach (GWR), the intercept model (using GWR package [10]) as well as a Poisson regression model as a function of latitudes and longitudes (using SAS [23]). The spatial autocorrelation 'best' model was estimated by the variogram (using GS+ [9]) by minimising the residual sum of squares (RSS) criteria. The second step was to explain the observed heterogeneity by the available covariables through regression models. Consider the general model:

$$g(E[y]) = X_i\beta + U_i$$

where y is the rate variable, g is the link function and $U_i, i = 1, 2, \ldots, m$ denotes the area-specific random effects (spatially unstructured and structured effects). A possible way to take into account spatial variability in a Poisson regression analysis of rates is to use the approach of Kleinschmidt et al. [11]. Their idea is the following: First, we suppose that there is no spatial autocorrelation. We then perform an ordinary Poisson regression model (with no correlation structure) as a function of the deprivation indices, calculate the signed residuals, and fit a theoretical semivariogram model (using GS+) to estimate the nugget σ_1^2, the sill σ^2 and the range ρ. These three parameters are then used to define a covariance matrix $\mathbf{V}$ associated with the spatial random effects:

$$\mathbf{V} = \mathbf{I}\sigma_1^2 + \mathbf{F}\sigma^2 \text{ with } F_{ij} = \exp(-d_{ij}/\rho)$$

were d_j refers to the distance between centroids of LHUs i and j, and $\mathbf{I}$ is an identity matrix. We assume here an exponential model for the spatial structure, but F_{ij} could be given by any other theoretical model adjusted to the variogram. Then, we fit a spatially correlated Poisson-mixture model (using the GLIMMIX SAS macro) with the covariance matrix $\mathbf{V}$ defined above, and calculate a second set of signed residuals. We iterate this process until there is convergence in the estimates. Another way of taking into account the spatial variability is to use a GWR model [8], the idea being to allow for geographical variation in the parameters β instead of fitting global regression models. The Gaussian model is given by:

$$y_i = \beta_0(u_i, v_i) + \sum_k \beta_k(u_i, v_i)x_{ik} + \varepsilon_i$$

where $(u_i v_i)$ denotes the coordinates of the ith LHU. If we assume that LHUs far from each other are more likely to have different coefficients, a weighted calibration $(\mathbf{W})$ is used, which is a Kernel-type function of the distance and a varying bandwidth. The coefficients β are thus estimated by:

$$\hat{\beta}(u_i, v_i) = (\mathbf{X}'\mathbf{W}(u_i, v_i)\mathbf{X})^{-1}\mathbf{X}'\mathbf{W}(u_i, v_i)\mathbf{y}$$

where $\hat{\beta}$ contains $k + 1$ continuous functions of geographical coordinates. Because the GWR package was not yet available for binary and count data in 2003, we modeled the logarithm of the rates, and used the normal approximation for the error term.

3 Results

A total of 50,839 patients met the inclusion criteria and were listed in the MED-ECHO register between January 1st 1996 and December 31st 1998. Almost 10% of the individuals ($n = 4,749$) died during the "index hospitalization". Thirty-three patients were excluded because of their age (< 25 years old) or because of an error in the code of residence. Furthermore, 1,516 patients had been hospitalized in the year preceding the index hospitalization

Table 1. ACS incidence and early readmission rates

	ACS incidence	non-fatal incidence	early readmission
women	17,841	15,205	1,305 (8.6%)
< 55 years	2,059	1,991	179 (9.0%)
≥ 55 years	15,782	13,214	1,126 (8.5%)
men	31,449	28,649	2,785 (9.7%)
< 45 years	2,419	2,389	247 (10.3%)
≥ 45 years	29,030	26,260	2,538 (9.7%)
total	49,290	43,854	4,090 (9.3%)

and were also excluded, for a total cohort of 49,290 patients. Among these, 43,854 (89%) were alive 30-days after discharge. At the index hospitalization, the average age of patients was 66.2 years (± 13.3), and men represented 63.8% ($n = 31,449$) of total cohort. A total of 4,090 early readmissions (9.3%) have been observed. The early readmission rate is higher for men but lower for older patients (Table 1). Figure 1 shows the 1996-population within LHU of the province of Quebec, while Figs. 2 and 3 show material and social deprivation indices respectively. We can easily see a large population density in the south part of the province, as well as in the coasts of the St-Laurent River. In addition, as measured with deprivation indices, we can observe that the urban regions are generally less deprived in the material sense but the reverse is observed for the social deprivation index.

Crude ACS incidence and readmission rates are shown in Figs. 4 and 5. The most likely clusters are highlighted as black lines. In Figs. 6 and 7 however, we present the smooth surface estimates of ACS incidence and early readmission rates with the GWR method (intercept model only; Monte Carlo test for spatial variability). The parameter estimates range from 0.00086 to 0.01266

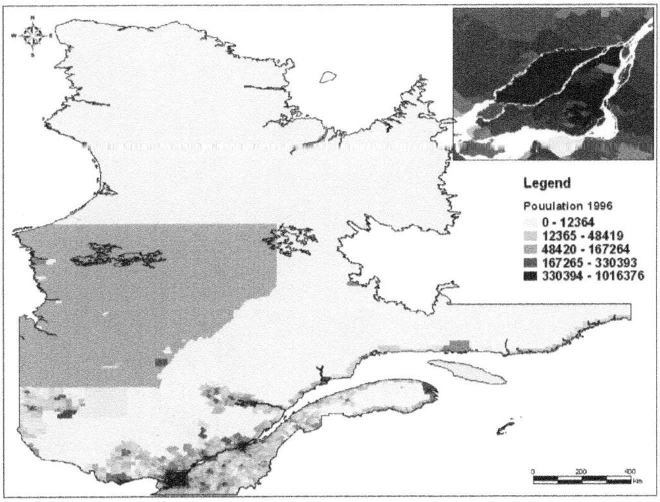

Fig. 1. Population of Quebec in 1996

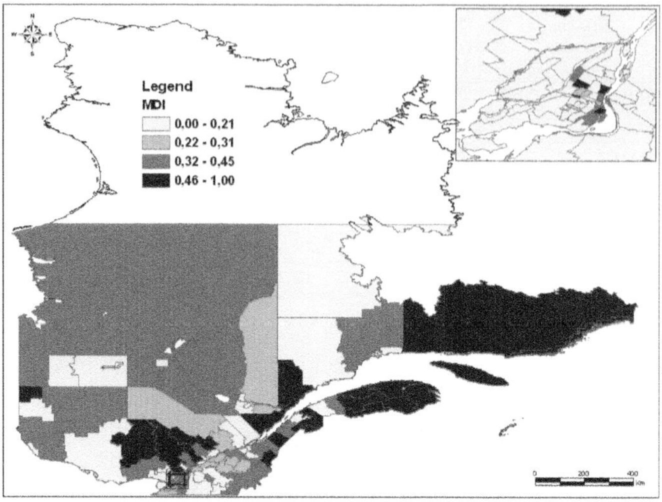

Fig. 2. Material Deprivation Index (MDI)

($p < 0.0001$) for ACS incidence and from 0.08093 to 0.13639 ($p = 0.0300$) for readmission rates. There is a North-East trend for ACS incidence rates while the readmission rates tend to increase as we move away from the urban centers (see smoothed surfaces) (Figs. 6 and 7).

Poisson regression models of the number of ACS and readmissions as a function of latitude and longitude showed similar trends (predicted rates not shown), confirming the observed trend from the GWR analysis. The method proposed by Kleinschmidt et al. [11] was explored but the semivariogram of the signed residuals suggested the absence of an autocorrelation structure, so

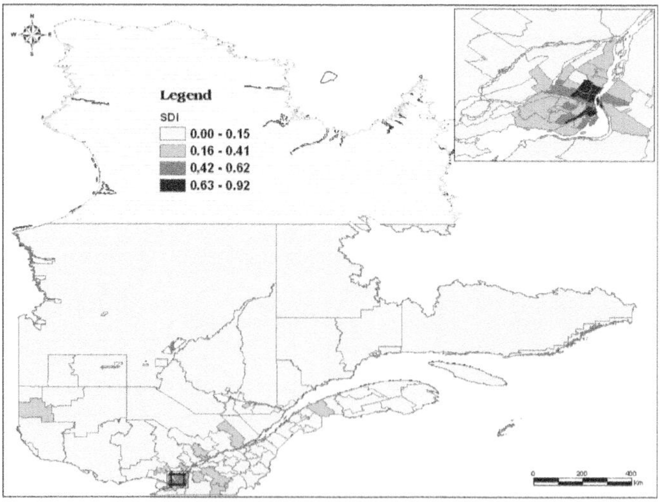

Fig. 3. Social Deprivation Index (SDI)

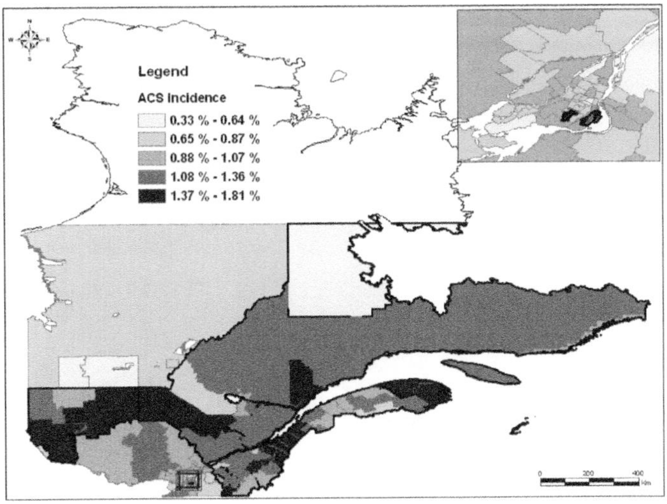

Fig. 4. Crude ACS incidence with hot spots (black lines)

the process stopped at the first iteration. Nevertheless, to see if the deprivation indices explain the heterogeneity in the rates, we performed a GWR on the logarithm of the rates as a function of these two indices. The parameter estimates for readmission rates were significant ($p < 0.0001$) while for ACS incidence rates only the SDI estimate was significant ($p < 0.0001$). The residuals were the ratio of observed and expected rates. Maps of surface residuals (Figs. 8 and 9) show that the two indices explain some of the variation but the observed trend in some regions remains unexplained by the deprivation indices.

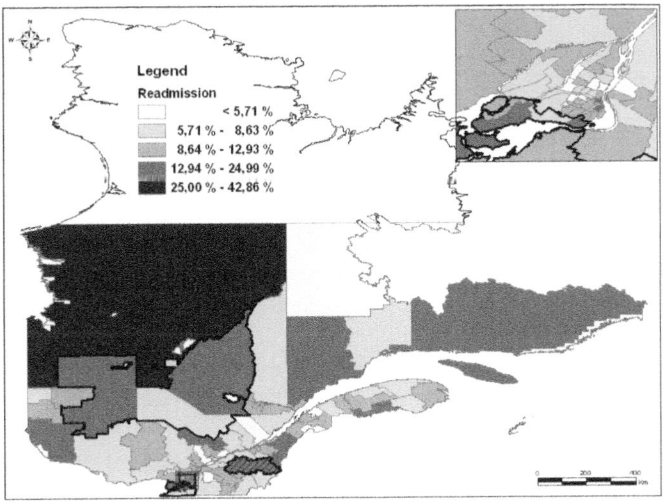

Fig. 5. Crude readmission rates with hot spots (black lines)

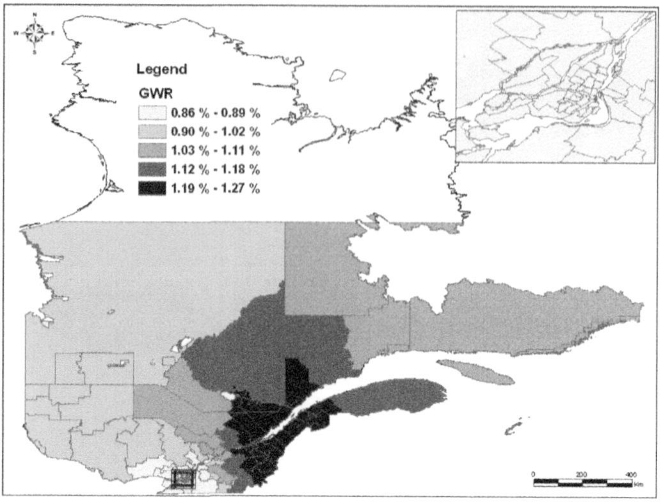

Fig. 6. Smoothed ACS incidence (GWR intercept estimates)

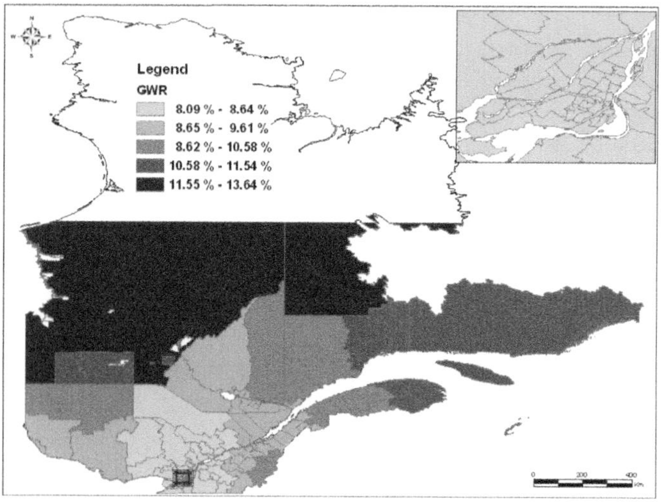

Fig. 7. Smoothed readmission rates (GWR intercept estimates)

4 Discussion

Spatial analysis, in the sense of analysis of data in a geographical perspective, has to be linked to geomatics and geographical information systems. Within the context of public health issues, a map displaying geographic heterogeneity is one of the most powerful tools for interpretation of spatial data once we determine which surface estimate best portrays the data and which estimation methods are appropriate for the health parameters of interest. Elliott et al. [6] reviewed estimation methods such as Kernel-based and GAM-based methods

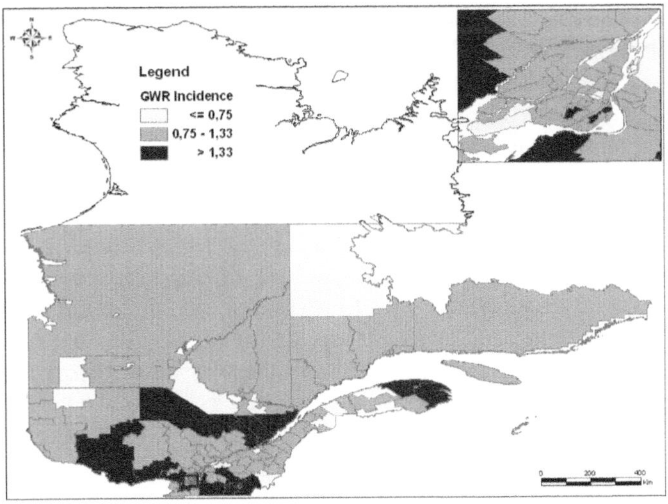

Fig. 8. ACS incidence as a function of MDI and SDI (GWR residuals)

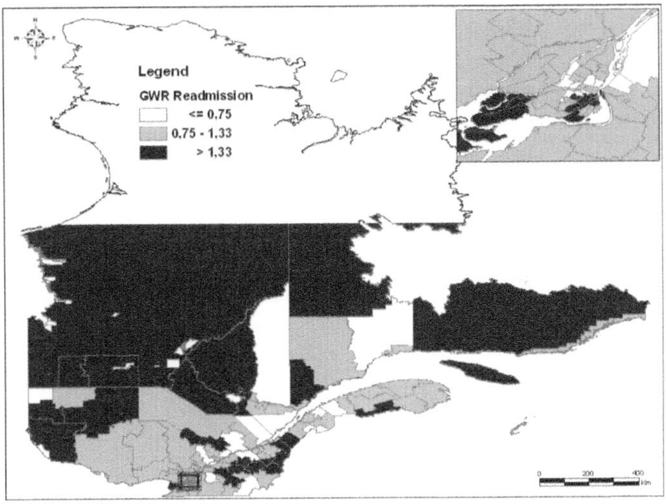

Fig. 9. Readmission rates as a function of MDI and SDI (GWR residuals)

as well as estimation methods involving the mixture of distributions. Lawson et al. [13] motivated the use of model-based estimation methods, especially methods involving random effects within a Bayesian framework. Cressie [4] pointed out the full potential for hierarchical spatial model-based methods using full Bayesian and Monte Carlo Markov Chain approaches. However, an important question for future consideration is what we gain through the use of additional complexity from the crude surface to Kernel-based to empirical Bayes to full Bayesian methods. Future work will also explore the added value of the hierarchical setting by using other patient and LHU level covariables as potential predictors with a hierarchy in the parameters of interest within the Bayesian framework and in the context of no prior information (flat priors) or limited information. Another promising avenue for exploration is how Monte Carlo simulations may be used to inform the choice of prior distributions. In the context of establishing relationships between health events and covariables (within exploratory or confirmatory spatial data analysis), it would be quite interesting to combine both Kernel-based methods such as the GWR approach and random effects models to deal simultaneously with large scale spatial variability (uncorrelated heterogeneity) and small scale variability (spatial autocorrelation). Another important consideration is the difference between semivariogram approach to model the spatial autocorrelation structure and conditional autoregressive (CAR) models [15]. As Fotheringham et al. [8] pointed out, the GWR approach offers an interesting framework by allowing estimation of a continuous surface for each potential correlate which could usefully be mapped. This approach offers a fertile ground for future study.

Acknowledgements

This project was subsidized by the Network of Centres of Excellence GEOIDE and Merck Frosst Canada Ltd. At the time of the study, the first author had the support of the Clinical Research Center of the Centre Hospitalier Universitaire de Sherbrooke.

References

1. ARCGIS. Release 8.2. ESRI, USA
2. Ashton CM, Kuykendall DH, Johnson ML, et al (1995) The association between the quality of inpatient care and early readmission. Ann Intern Med 122:415–21
3. Cardiologie tertiaire, situation actuelle, perspectives et propositions, 2000. Ministère de la Santé et des Services Sociaux
4. Cressie N (2000). Spatial statistics and environmental sciences. In: proceedings of the section of statistics and the environment. American Statistical Association. Alexandria, VA,1–10
5. DMTI Spatial Data Delivery System (2000) CanMap Streetfiles et PostCode files, http://www.dmtispatial.com
6. Elliott P, Wakefield JC, Best NG, Briggs DJ (2000) Spatial Epidemiology: Methods and Applications. Oxford University Press, New York
7. ESRI Date & Maps 2002, Media Kit, 2002, ESRI ArcGIS, http://www.esri.com
8. Fotheringham AS, Brunsdon C, Charlton M (2002) Geographically Weighted Regression: The Analysis of Spatially Varying Relationships. John Wiley & Sons, England
9. GS+ Geostatistics for the Environmental Sciences, version 5.3.2. γ Design Software. Painwell, MI, USA
10. GWR Version2.0.5. University of Newcastle, England
11. Kleinschmidt I, Sharp BL, Clarke GPY, Curtis B, Fraser C (2001) Use of generalized linear mixed models in the spatial analysis of small-area malaria incidence rates in KwaZulu Natal, South Africa. Am J Epidemiol 153(12):1213–21
12. Kulldorff M (1997) A spatial scan statistic. Commun. Stat.: Theory Methods,26:1481–96.
13. Lawson AB, Browne WJ, Vidal Rodeiro CL (2003) Disease Mapping with WinBUGS and MLwiN. Wiley & sons, New York
14. Levy AR, Tamblyn RM, Fitchett D, McLeod PJ, Hanley JA (1999) Coding accuracy of hospital discharge data for elderly survivors of myocardial infarction. Can J Cardiol 15(11):1277–82
15. McNab Y, Dean CB (2000) Parametric bootstrap and penalyzed quasi-likelihood inference in conditional autoregressive models. Stat. Med., 19(17–18), 2421–35
16. Mehta RH, Montoye CK, Gallogly M, et al for the GAP Steering Committee of the American College of Cardiology (2002) Improving quality of care for

acute myocardial infarction: the guidelines applied in practice (GAP) initiative. JAMA 287(10):1269–76
17. Ministère de la Santé et des Services Sociaux, http://www.msss.gouv.qc.ca
18. Pampalon R, Raymond G (2000) Un indice de défavorisation pour la planification de la santé et du bien-être au Québec. Maladies Chroniques au Canada 21(3):113–22
19. Petersen LA, Wright SM, Normand SLT, Daley J (1999) Positive predictive value of the diagnosis of acute myocardial infarction in an administrative database. J Gen Intern Med 14:555–8
20. Pilote L, Califf RM, Sapp S, et al (1995) Regional variation across the United States in the management of acute myocardial infarction. GUSTO-1 Investigators. Global utilization of streptokinase and tissue plasminogen activator for occluded coronary arteries. N Engl J Med 333(9):565–72
21. Pilote L, Granger C, Armstrong PW, et al (1995) Differences in the treatment of myocardial infarction between the United States and Canada. A survey of physicians in the GUSTO trial. Med Care 33(6):598–610
22. Ryan TJ, Antman EM, Brooks NH, et al (1999) 1999 update: ACC/AHA guidelines for the management of patients with acute myocardial infarction. A report of the ACC /AHA Task Force on practice guidelines. J Am Coll Cardiol 34(3):890–911
23. The SAS systems for windows. Release 8.02. SAS Institute Inc. Cary, NC, USA
24. Thomas JW, Holloway JJ (1991) Investigating early readmission as an indicator for quality of care studies. Med. Care 29:377–94
25. Van der Werf F, Topol E, Lee K, et al (1995) Variations in patient treatment and outcomes for acute myocardial infarction in the United States and other countries. JAMA 273:1586–91
26. Weissman JS, Ayanian JZ, Chasan-Taber S, et al (1999) Hospital readmissions and quality of care. Med. Care 37:490–501
27. Wenneberg JE (1984) Dealing with medical practice variations: a proposal for action. Health Aff (Millwood) 3:6–32
28. Williams JI, Young W (1996) Appendix – A Summary of Studies on the Quality of Health Care Administrative Databases in Canada. In: Goel V, Williams JI, Anderson GM, Blackstein-Hirsch P, Fooks C, Naylor CD (eds) Patterns of Health Care in Ontario : The ICES Practice Atlas. 2nd ed. Ottawa, Ontario : Canadian Medical Association, pp 339–45

Interactive Cumulative Curves for Exploratory Classification Maps

Gennady Andrienko and Natalia Andrienko

Fraunhofer Institute AIS Schloss Birlinghoven, Sankt Augustin, Germany
gennady.andrienko@ais.fraunhofer.de

1 Classification as an Instrument for Exploratory Analysis

Representation of values of a numeric attribute referring to geographical objects, in particular, areas of territory division, is often done in cartography using the technique of classification. According to this technique, the value range of the attribute is divided into intervals. Then different colours are chosen to represent values from each of the intervals on a map.

Historically, classification was indispensable due to technical limitations involved in production of paper maps as well as in display of maps on early graphical computer screens [6]. With appearance of modern computer hardware these limitations were eliminated, and it became possible to produce unclassed maps. In such maps numeric values are encoded by proportional degrees of darkness.

The merits and flaws of classed and unclassed maps have long been a topic of hot debates in cartography. It is not our intention to recite here this discussion or to take either side in it. Although classed maps are in the focus of this paper, this does not mean that we regard them as superior to unclassed maps. Our interest is the use of maps in exploratory analysis of spatially referenced data. In such analysis each type of maps plays its own role, and, hence, there is no question whether to select a classed or an unclassed map. An analyst should use both because these are complementary instruments of analysis.

Our opinion can be substantiated as follows. The goal of exploratory analysis is to gain understanding of given data, that is, to derive a short (compressed) description of their essential characteristics. Thus, according to Bertin, understanding is "discovering combinational elements which are less numerous than the initial elements yet capable of describing all the information in a simpler form" ([3], p. 166). With regard to spatial distribution of values of a numeric attribute, an analyst initially has one value per each spatial object. The description of this data set could be substantially simplified if there were a directional trend in value distribution, for example, increase of values from the north to the south or from the centre of the territory to its periphery. Alternatively, a shorter description could be derived if the territory could be divided into possibly smaller number of coherent regions with low variation of attribute values within the regions. This technique is known as regionalisation. Unclassed maps are better suited for detecting trends because they do not hide differences. Classification discards differences between values within a class interval and gives the corresponding objects similar appearance on the map. When these objects are geographical neighbours, they tend to be visually associated into clusters. This property makes classed maps well suitable for regionalisation. Which of the two ways to simplification occurs to be possible or more effective in each specific case, depends on the data and not on the preferences of the analyst. Therefore it is necessary to have both an unclassed choropleth map and a classification tool in order to investigate properly data with previously unknown characteristics.

It is clear, however, that a single static classed map cannot appropriately support regionalisation. It is well known in cartography that different selection of the number of classes and class breaks may radically change the spatial pattern perceived from the map [5, 7]. There is no universal recipe of how to get an "ideal" classification with understandable class breaks, on the one hand, and interpretable coherent regions, on the other hand. Therefore when we say that classification may be used as an instrument of data analysis, we mean not a classed map by itself but an interactive tool that allows the analyst to change the classes and to observe immediately the effect on the map.

The exploratory value of classification was recognised in cartography only relatively recently. Initially classification was regarded as a tool for conveying specific messages from the map author to map consumers. Thus, the paper [8] considers various possible intentions of the map designer and demonstrates how they can be fulfilled through application of different classification methods and selection of the number of classes.

In early nineties Egbert and Slocum developed a software system called ExploreMap intended to support exploration of data with the use of classed choropleth maps [4]. The most important feature of the system was a possibility to interactively change the classes. Another implementation of this function based on direct manipulation techniques is the "dynamic classification" tool incorporated in the system CommonGIS [1, 2]. Exploration on the basis of classification is additionally supported in CommonGIS by the function of computing statistics for the classes: the range of variation and the average, median and quartile values of any selected attribute for each class.

In this paper we describe a recently developed extension of the dynamic classification tool that exploits the properties of the cumulative frequency curve and generalised cumulative curves. In the next section we define the relevant notions and explain the use of cumulative curves in classification and data exploration.

2 Cumulative Frequency Curve and Its Use in Classification

In classification of spatially referenced data an analyst needs to consider the data from two perspectives, statistical and spatial, and take into account the peculiarities of both the statistical and the spatial distributions of the data. This means that the analyst needs to pursue at least two concurrent goals. The first is to minimise variation of data within each class and to maximise differences between classes. The second goal is to divide the territory into the smallest possible number of coherent regions with low data variation within the regions. Additional goals may emerge in particular application domains. Thus, in demographic applications it may be necessary to minimise differences between the classes in total population or total area. The analyst needs such tools that would allow her/him to balance between these goals in search of an acceptable compromise solution.

A visual representation of the statistical value distribution can help the analyst to meet the statistical criteria. Yamahira et al. [8] suggested that classification could be supported by a frequency histogram. However, a histogram represents results of prior classification and therefore can hardly serve as a tool for it. The dynamic classification tool of CommonGIS includes a dot plot, or point graph. This kind of graph does not require prior classification and is well suited for including in the interactive classification device due to its modest requirements to screen space. On the other hand, an obvious problem is overlapping of point symbols that obscures the understanding of the distribution. Slocum [7] uses so called dispersion graphs (for illustration purposes) in his discussion of the existing classification methods. This representation is based on classification into a large number of classes. Values fitting in a class are shown by dots stacked at the class position. Since dispersion graphs already involve classification, they are not so good as tools for producing other classifications.

One more method for graphical representation of statistical distribution is the cumulative frequency curve, or ogive. In such a graph the horizontal axis represents the value range of an attribute. The vertical position of each point of the curve corresponds to the number of objects with values of the attribute being less than or equal to the value represented by the horizontal position of this point. The method of construction of the ogive is demonstrated in Fig. 1. The curve represents the distribution of values of the attribute "Number of cars per capita" over districts of Leicestershire (this and further examples refer to the Leicestershire sample of the 1991 census data available at the URL http://www.mimas.ac.uk/descartes/).

Peculiarities of value distribution can be perceived from the shape of the ogive. Steep segments correspond to clusters of close values. The height of such a segment shows the number of the close values. Horizontal segments correspond to "natural breaks" in the sequence of values.

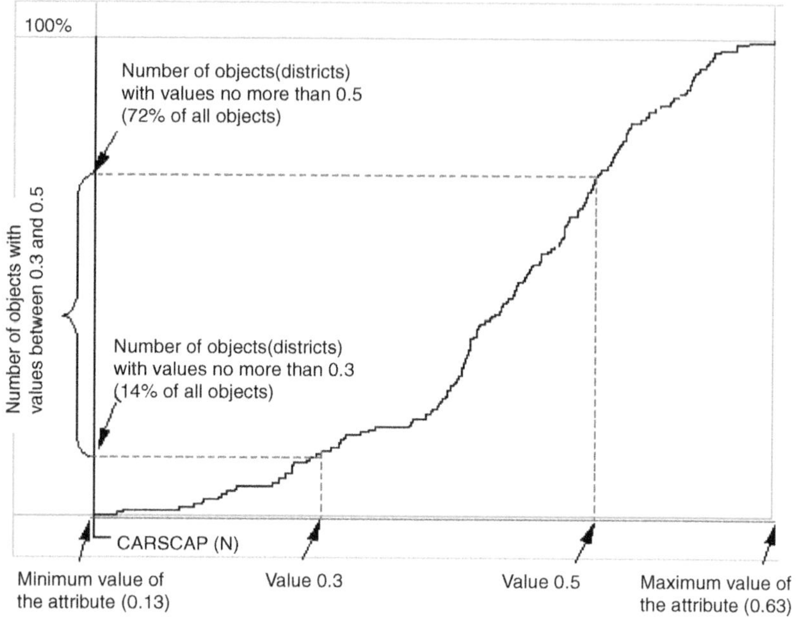

Fig. 1. A cumulative frequency curve

It is important that the cumulative frequency curve does not require prior classification. However, it can represent results of classification by means of additional graphical elements, and we used this opportunity in the latest extension of CommonGIS. Thus, the horizontal axis of the graph may be suited to show class breaks. In the interface adopted in CommonGIS (Fig. 2) we use for this purpose segmented bars with segments representing the classification intervals. The segments are painted in the colours of the classes. The positions of the breaks are projected onto the curve, and the corresponding points of the curve are, in their turn, projected onto the vertical axis. The division of the vertical axis is also shown with the use of coloured segmented bars. The lengths of the segments are proportional to the numbers of objects in the corresponding classes. With such a construction it becomes easy to compare the sizes of the classes. For example, the class breaks shown in Fig. 2 divide the whole set of objects into 3 groups of approximately equal size that is demonstrated by the equal lengths of the bar segments on the vertical axis.

The overall interface for classification is shown in Fig. 3. Besides the cumulative curve display exposing statistical characteristics of the current classification, it includes a map showing geographical distribution of the classes. For the use of the cumulative curve as a tool for classification it is important that its display immediately reacts to any changes of the classes, as well as the map does. The user can change class breaks by moving the sliders (double-ended vertical arrows) along the slider bar (on the upper right of the window). In

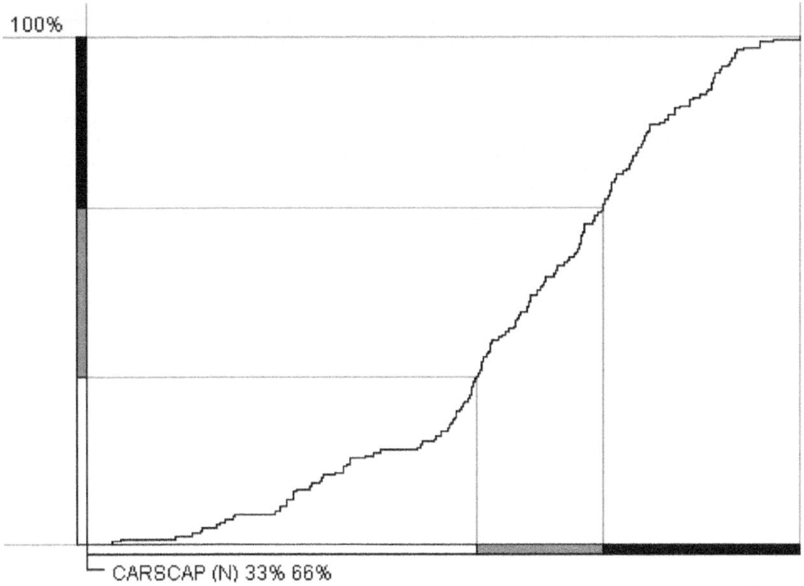

Fig. 2. Representation of classes on a cumulative frequency curve display. Values of the attribute "Number of cars per capita" are divided by 2 breaks (0.404 and 0.491) into 3 classes with approximately equal sizes (33, 33, and 34% of the whole set). The positions of the breaks on the vertical axis are indicated below the graph: 33 and 66% (of the total number of the classified objects)

the process of moving the slider the map and the cumulative curve graph are dynamically redrawn. In particular, changing are the relative lengths of the coloured bars on the axes and the positions of the projection lines. Clicking on the slider bar introduces a new class break, bringing a slider close to another slider results in the corresponding break being removed. The map and the cumulative curve display immediately reflect all these changes.

It is important that the cumulative frequency curve does not require prior classification. However, it can represent results of classification by means of additional graphical elements, and we used this opportunity in the latest extension of CommonGIS. Thus, the horizontal axis of the graph may be suited to show class breaks. In the interface adopted in CommonGIS (Fig. 2) we use for this purpose segmented bars with segments representing the classification intervals. The segments are painted in the colours of the classes. The positions of the breaks are projected onto the curve, and the corresponding points of the curve are, in their turn, projected onto the vertical axis. The division of the vertical axis is also shown with the use of coloured segmented bars. The lengths of the segments are proportional to the numbers of objects in the corresponding classes. With such a construction it becomes easy to compare the sizes of the classes. For example, the class breaks shown in Fig. 2 divide

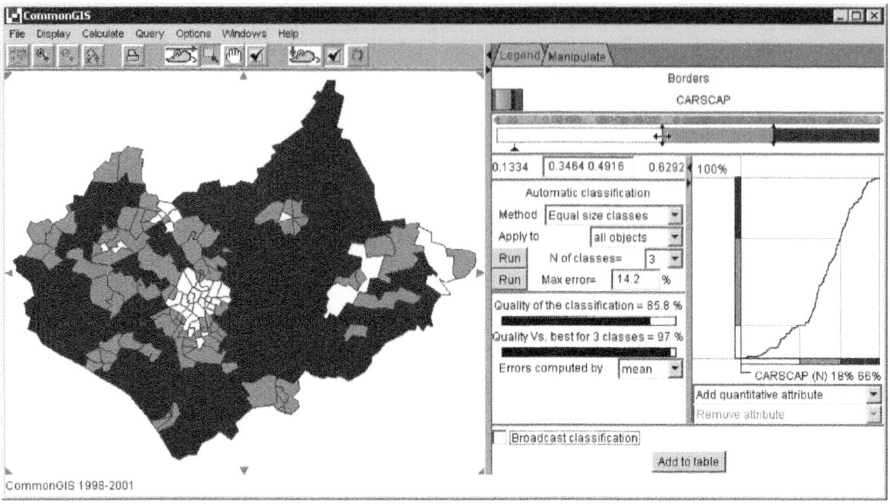

Fig. 3. The interface for classification provided in CommonGIS allows the user to account both for statistical and for spatial distribution of values

the whole set of objects into 3 groups of approximately equal size that is demonstrated by the equal lengths of the bar segments on the vertical axis.

The overall interface for classification is shown in Fig. 3. Besides the cumulative curve display exposing statistical characteristics of the current classification, it includes a map showing geographical distribution of the classes. For the use of the cumulative curve as a tool for classification it is important that its display immediately reacts to any changes of the classes, as well as the map does. The user can change class breaks by moving the sliders (double-ended vertical arrows) along the slider bar (on the upper right of the window). In the process of moving the slider the map and the cumulative curve graph are dynamically redrawn. In particular, changing are the relative lengths of the coloured bars on the axes and the positions of the projection lines. Clicking on the slider bar introduces a new class break, bringing a slider close to another slider results in the corresponding break being removed. The map and the cumulative curve display immediately reflect all these changes.

3 Generalised Cumulative Curves

In particular application domains additional classification criteria may come into play in combination with the statistical and geographical ones. For example, in demographic analyses it may be important to produce classes of districts that do not differ too much in total number of population that lives in them.

It is possible to generalise the idea of the cumulative frequency curve and to build similar graphs summarising values of arbitrary quantitative attributes. Examples of such attributes are area, population number, gross domestic product, number of households, etc. A generalised cumulative curve is built in the following way. Let the horizontal axis correspond to attribute A and the vertical to attribute B. The curve matches each value x of A with the sum of values of B computed for objects with the values of A being less than or equal to x. So, the maximum value of A will correspond to the total sum of values of B for all the objects of the sample. Let, for example, the districts of Leicestershire be classified according to the number of cars per capita, and a generalised cumulative curve represent the attribute "Total population". Then the vertical position corresponding to x cars per capita would reflect the total number of population living in districts with no more than x cars per capita.

The classification tool of CommonGIS allows the user to add a generalised curve for any quantitative attribute to the cumulative frequency curve display. The curves are overlaid, i.e. drawn in the same panel (see Fig. 4). To be better discriminated, the curves differ in colour. The horizontal axis is common for all of them. The vertical axes are shown beside each other on the left of the graph. Each of the vertical axes is divided into the same number of segments, but positions of the breaks are, in general, different. This is clearly demonstrated in Fig. 4. It shows that 34% of all districts fit in the lower class of the classification. They occupy only 10% of the total area but contain 50% of total population.

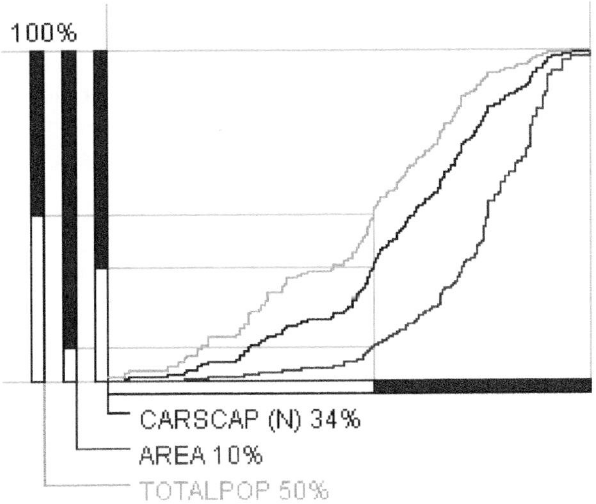

Fig. 4. Generalised cumulative curves are built for the attributes "Area" and "Total population". The classification is done on the basis of the attribute "Number of cars per capita"

Having such a tool, it is easy to account in classification for such criteria as even distribution of population among the classes, or approximately equal total areas occupied by the classes, or other specific criteria that may emerge in this or that application domain. Thus, to make classes approximately equal in total population, the analyst should focus in the process of slider movement on the axis corresponding to total population and try to position the sliders so that the axis is divided into segments of equal length.

Besides this opportunity, generalised cumulative curves may be used for exploring relationships between various characteristics of the classified objects. Let us demonstrate this on the example of exploration of unemployment in Leicestershire. We used the attributes "Number of unemployed" and "Total population" to calculate percentage of unemployed in total population in each district. Then we took this new attribute as the basis for classification. The classification tool showed us that proportion of unemployed in population varies from 0.9 to 13.62%. We considered values above 4% to be very high and wondered in how many districts this threshold is exceeded and where these districts are located. We entered 4 as a class break and in this way divided all districts into two classes: with up to 4% of unemployed persons in population and with more than 4%. The cumulative curve display showed us that only 18% of all districts fit in the upper class (Fig. 5). The map shows a vivid spatial cluster of such districts in the centre of the area. It is seen that these districts occupy a rather small part of the whole area. However, when we selected the attribute "Total population" for representation on the cumulative curve display, we found that the districts with high unemployment contain 33% of the total population of Leicestershire.

We became interested whether there is a link between unemployment and distribution of national minorities. We added to the cumulative curve display the attribute BLAFR representing total numbers of people originating from Africa by districts. In Fig. 5 it is vividly seen that the curve for this population group radically differs from that for the whole population. It is also seen

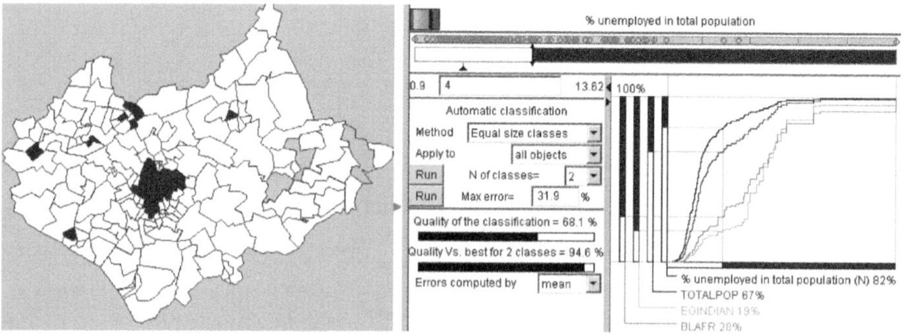

Fig. 5. The use of generalised cumulative curves for exploration of unemployment in Leicestershire

that the axis corresponding to this attribute is divided in quite a different proportion than those for the frequency and for the whole population. Only 28% of people with African origin live in districts with lower unemployment and, hence, 72% live in districts with more than 4% unemployed in total population. The difference is even more dramatic for people originating from India (represented by the attribute EGINDIAN). One can see that 81% of these people live in the areas with high unemployment.

We continued our investigation of unemployment by moving the class break so that the population was divided into two equal parts. Figure 6 shows the result of this operation. The new value of the class break is 2.89. This means that 50% of total population lives in districts with more than 2.89% of unemployed. These districts, as it is seen from the map, constitute a rather small part of the whole territory of the county. Hence, the population density in them is higher than in the rest of the districts. Apparently, these are mainly urban districts. The map shows also that the districts with higher unemployment are spatially clustered. The national minorities considered above are now distributed between the classes of districts in the following way: only 14% of Indians and 18% of Africans live in the districts with lower unemployment, and, hence, 86 and 82%, respectively, live in the areas with more than 2.89% unemployed in population.

Classification with multiple criteria involved but also to reveal significant relationships in data. However, it should be borne in mind that this technique is suitable only for attributes the values of which can be summed up over the set of objects they refer to. For example, it would be wrong to apply it to percentages, averaged values, rates, values per capita etc.

The implementation of the interactive classification tool based on the use of cumulative curves is done in Java. This allows the use of it on different platforms and in the WWW. The new version of the system CommonGIS that includes the tool described can be run in the WWW at our homepage.

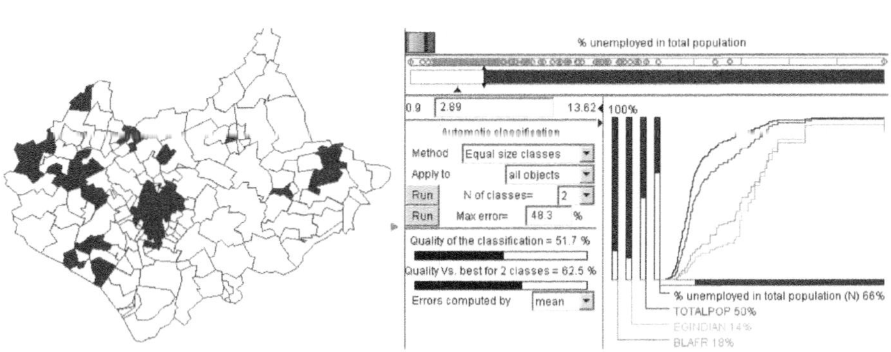

Fig. 6. The cumulative curve display was used to divide the districts into two classes with equal total population

4 Conclusion

Interactive, dynamic classification can be a valuable instrument in exploratory analysis of spatially referenced data. Representation of classes on a map by colouring facilitates perception of patterns of spatial distribution. Therefore we included a dynamic classification tool into the array of exploratory facilities offered by our system CommonGIS.

In classification of geographical objects according to values of a numeric attribute it is necessary to take into account peculiarities of both spatial and statistical distributions of values. In search for a suitable representation of the statistical distribution we studied the properties of the cumulative frequency curve and found it to be a good solution. This representation allows a two-way use. On the one hand, one can visually evaluate statistical characteristics of a given classification. On the other hand, one can produce classifications with desired statistical characteristics. In our implementation a cumulative curve display is included in the interface for classification together with direct manipulation controls and a map showing the results of classification in the geographical space. In this interface the user can gradually shift class breaks and immediately observe the effect on the map and on the cumulative curve display. This dynamic link between the components of the interface allows the user to evaluate "on the fly" lots of variants of classification from the perspective of satisfaction of geographical and statistical criteria and to arrive eventually at a good compromise solution.

In the process of exploration of the properties of the cumulative frequency curve we came upon an idea that this representation could be extended to arbitrary attributes that allow summing over a set of objects. Just as the frequency curve accumulates the number of objects, the generalised curve would accumulate the values of an attribute. The use of such curves in classification offers additional opportunities. One of them is accounting in classification for such criteria as even distribution of population or area among classes. Another opportunity is investigation of relationships between the attribute used for classification and various quantitative characteristics of the objects being classified.

We believe that inclusion of cumulative curves in the tool for classification available in CommonGIS significantly increases its exploratory value. Thus, this tool has got a high appraisal of an expert in statistics and statistical graphics and a professional in analysis of geographic information, also with a solid statistical background. However, we have a concern that the users with less expertise in statistics may find cumulative curves difficult to understand. We plan to perform experiments in order to evaluate how people can handle the classification tool and how much time they need to comprehend the cumulative curve display and learn to use it.

References

1. Andrienko, G. and Andrienko, N. (1998) Dynamic Categorization for Visual Study of Spatial Information. Programming and Computer Software, 24 (3), 108–115.
2. Andrienko, G. and Andrienko, N. (1999) Interactive maps for visual data exploration. International Journal Geographical Information Science, 13 (4), 355–374.
3. Bertin, J. (1965/1983) Semiology of graphics. Diagrams, networks, maps. The University of Wisconsin Press, Madison WI.
4. Egbert, S.L. and Slocum, T.A. (1992) EXPLOREMAP: an exploration system for choropleth maps. Annals of the Association of American Geographers, 82, 275–288.
5. MacEachren, A.M. (1994) Some truth with maps: a Primer on symbolization and design. Association of American Geographers, Washington.
6. Robinson, A.H., Morrison, J.L., Muehrcke, P.C., Jon Kimerling, A., and Guptil, S.C. (1995) Elements of cartography. Wiley, New York.
7. Slocum, T.A. (1999) Thematic cartography and visualization. Prentice-Hall, New Jersey.
8. Yamahira, T., Kasahara, Y., and Tsurutani, T. (1985) How map designers can represent their ideas in thematic maps. The Visual Computer, 1, 174–184.

Combining REmbeddedPostgres and PostGIS

Albrecht Gebhardt

Department of Statistics, University of Klagenfurt, Klagenfurt, Austria
agebhard@uni-klu.ac.at

1 Introduction

This work presents some results of combining several pieces of Open Source software, including R[1](see [6]), PostgreSQL[2](see [2]), and some appropriate extension packages, see [3]. A central building block is REmbeddedPostgres, an extension of the PostgreSQL RDBMS, which basicly delivers the possibility to use R syntax within SQL queries. Using the work of Duncan Temple Lang [4] as starting point, the idea arose, to implement more complex statistical database queries. Finally the focus has been put on queries involving spatial data. At this point PostGIS comes into action. It is another PostgreSQL extension and implements OpenGIS functions. This extension enables PostgreSQL to process spatially referenced data.

An implementation of a "linear model" query will be shown. It involves several modifications of REmbeddedPostgres and needs some extra SQL functions written in R and Perl. This can easily be combined with OpenGIS SQL functions. As a result a "spatial statistical SQL query" becomes possible.

2 GIS, RDBMS and Statistical Software

GIS, database systems and statistical software fulfill different tasks in the analysis of spatial data. GIS are used for collecting and editing data, generation of maps, transformation of and operations with maps. Database systems hold data, are used for indexing and selection by queries and can combine different portions of data via joins between tables. Finally the tasks of statistical software are exploratory analysis, modelling and estimation.

Several combinations of GIS, database systems (DBMS) and statistics software are possible:

---
[1] `http://www.r-project.org/`
[2] `http://www.postgresql.org/`

- DBMS as database engine for a GIS
- Database interfaces for statistical software or GIS
- GIS interfaces for statistical software
- statistics extensions for GIS

Figure 1 tries to summarize those combinations for R, PostgreSQL and the Open Source GIS GRASS.[3] R can interact with both GRASS and PostgreSQL by means of appropriate R libraries. GRASS GIS can access data stored in a PostgreSQL database. PostgreSQL delivers GIS related functions via its PostGIS extension and it can provide statistical functions via an embedded R interpreter. During the following sections we will only focus on the relations between R and PostgreSQL.

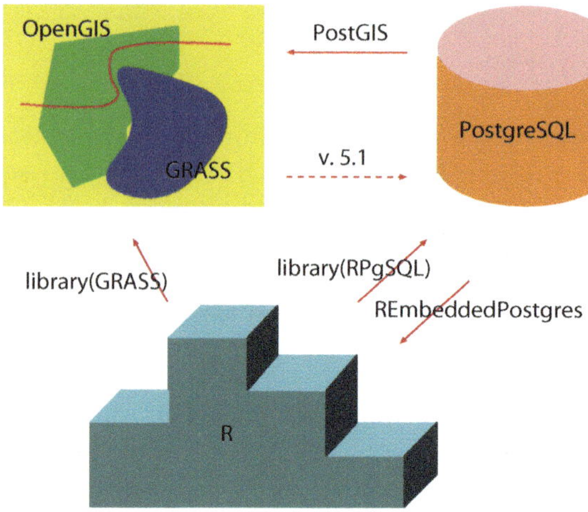

Fig. 1. GRASS, R and PostgreSQL

3 PostgreSQL

PostgreSQL is an open source object relational database server. It is freely available under a BSD style licence. Its roots reach back to 1977 where the developement of INGRES started as a research project at Berkeley University, California. In 1986 Michael Stonebreaker started a new project based on INGRES (which had been converted into a commercial product) and called it Postgres. Meanwhile it has been redesigned at least two times and changed its name via Postgres95 to PostgreSQL in 1996. At this time the query language of Postgres was changed to SQL. Now PostgreSQL almost completely implements the ANSI SQL/92 standard. Ordinary queries have the form

[3] `http://grass.itc.it`

```
SELECT x,y,z FROM table WHERE condition;
```

PostgreSQL is available for most UNIX/Linux systems. Also a native port for Win32 systems exists. One of the advantages is the extensibility of PostgreSQL. A first possibility to extend PostgreSQL is the use of its PL/PGSQL procedural language. This makes it easy to develop new database functions using the SQL language. But it is also possible to write C code and to make these C functions available in SQL as a new procedural language. This is the way how PostGIS and REmbeddedSQL have been implemented.

4 PostGIS

PostGIS is an OpenGIS implementation for PostgreSQL and follows the "Simple Features Specification for SQL",[4] see [5] . The OpenGIS[5] standard defines a minimum set of geometric data types, operations and functions for a database to become useable as backend for a GIS. Most commercial databases contain such OpenGIS extensions, examples are Oracle Spatial, IBM DB2 Spatial Extender and Informix Spatial DataBlade.

PostGIS implements additional data types like point, line, polygon etc. It stores geometry information in a so called geometry table. New functions like "Distance", "Area2d", "Box3D", to mention just a few of them, are now available. A full list of functions and detailed introduction can be found in the online documentation[6] of PostGIS.

PostGIS functions can be used to extend ordinary SQL queries to "geometric" queries:

```
SELECT x,y,z FROM table
WHERE Distance(GeomFromText('POINT(x0 y0))',
               SRID(geopoint)), geopoint)<r;
```

In this simple example we would select all values x, y, z within a circle $(x, y)^\top \in B((x_0, y_0), r) \subset \mathbb{R}^2$.

5 REmbeddedPostgres

REmbeddedPostgres[7] is part of the omegahat[8] project. The development of REmbeddedPostgres was driven mainly by the two ideas. Running an R interpreter within the database saves much data transmission time because computation takes place at the server and not at a database client. Additionally

[4] http://www.opengis.org/techno/specs/99-049.pdf
[5] http://www.opengis.org/
[6] http://postgis.refractions.net/docs/
[7] http://www.omegahat.org/RSPostgres/
[8] http://www.omegahat.org

database users could easily adopt to a statistical language if it is following the SQL syntax. Reference [4] gives a technical overview and discusses examples like prediction using an imported linear model.

REmbeddedPostgres implements both record and aggregate functions (also referred to as PL/R functions). Record functions can be used to apply simple R scripts to several arguments. Aggregate functions are more complex. They determine the result iterative by presenting the data sequentially to an "update function" which saves its results in a state variable.

6 Combination

6.1 Simple Queries

After installing both components into a PostgreSQL server it is easy to build simple "spatial statistical" SQL queries. E.g. if the mean of some variable of interest within some circular region is to be retrieved one could use

```
SELECT mean(z) FROM table
WHERE Distance(GeomFromText('POINT (x0 y0))',
               SRID(geopoint)), geopoint)<r;
```

Other examples can be created easily by combining different R functions for univariate statistics with different OpenGIS geometry functions.

6.2 More Advanced Functions

It would also be desirable to have the possibility to execute more complex queries like

```
SELECT lm('z~x+y') FROM table
WHERE Distance(GeomFromText('POINT(x0 y0))',
               SRID(geopoint)), geopoint)<r;
```

Because SQL aggregate functions can only be applied to a single argument we have to introduce new data types which combine the arguments x, y, z into a vector $(x, y, z)^\top$. If we restrict the linear model to multiple linear regression, we can implement this using user defined float8 based vector types. In this case it is necessary that REmbeddedPostgres can handle these new data types and has access to appropriate type conversion routines.

7 Technical Details

The idea is to extend REmbeddedPostgres in the following way to be able to handle user defined data types:

- Create user defined types (needs input and output functions written in C for PostgreSQL internal use).
- Write type conversion routines similar to the conversion routines for float8 already contained in REmbeddedPostgres.
- Introduce an additional table `repg_utypes` which contains the type OID, type name, the names of to/from R converter functions and pathname of a shared object file containing the shared library code which implements the converter functions.

It is also necessary to add a "user type registration function" to the R initialization in REmbeddedPostgres which reads the table `repg_utypes`, dynamically loads the shared library, accesses the converter routines (via the `dlopen` call) and adds registration info to the converter routines structure.

A first prototype of this approach is working. It reads the table `repg_utypes` within the C code by connecting back to the data base engine using the libpq C interface.

Current development focuses on improving the above mentioned REmbeddedPostgres extension. This makes use of PL/PERL as another PostgreSQL extension module, which implements an internal interface to the Perl[9] scripting language. Combining PL/R and PL/PERL can help in simplifying the notation of more complex PL/R queries.

The following problem arises. User data types have to be of a fixed length. That means different types for several vector dimensions together with several converting functions have to be written. This leads to a more complex syntax of the PL/R queries, because calls to the type conversion routines have to be added. Using PL/PERL can help here to hide these details from the user. Perl scripts analyze the arguments and construct the PL/R queries. The appropriate vectorized query can than be executed by using the DBD::PgSPI[10] Perl module for internal database access from PL/PERL (see [1]).

Currently we can apply e.g. R's linear model function `lm()` to a spatial subset selected by means of OpenGIS functions in two steps. First creating a SQL view and then applying the PL/PERL function `lm` to this view. The PL/PERL function performs parsing of the linear model formula given in S notation and then calls appropriate type conversion routines and finally executes the PL/R aggregate function containing the call to the R function `lm()`.

The result of the following query would be the estimated parameter vector θ of a linear model $z = \theta_0 + \theta_1\, x + \theta_2\, y + \varepsilon$.

```
CREATE VIEW spatial_view AS
SELECT X(geopoint),Y(geopoint),z FROM table
WHERE Distance(GeomFromText('POINT (x0 y0))',
               SRID(geopoint)), geopoint)<r;
SELECT lm('z~x+y','spatial_view');
```

[9] `http://www.perl.org/`

[10] `http://jamesthornton.com/postgres/7.3/postgres/plperl-database.html`

This application is mainly based on the open programming interface of PostgreSQL, which made it possible that several PostgreSQL extensions exist: an OpenGIS implementation, an embedded R interpreter within the database and a Perl interface. The combination of these building blocks made it possible to implement some "spatial statistical" functions directly within the database server by using synergy effects. The fact that all parts consist of freely available software is also very important.

References

1. A. Descartes and T. Bunce (2000) *Programming the Perl DBI.* O'Reilly & Associates, Inc.
2. J. Hartwig (2001) *PostgreSQL Professionell und praxisnah.* Addison-Wesley.
3. K. Hörhan (2002) Integration von Statistiksoftware in Datenbanksystemen und deren Anwendung in der räumlichen Statistik. Thesis, University of Klagenfurt
4. D. T. Lang (2001) Scenarios for using R within a relational database management system server. Technical report, Bell Labs.
5. Open GIS Consortium Inc. *The OpenGIS abstract specification* (1999). `http://www.opengis.org/techno/abstract.htm`
6. R Development Core Team (2004) *R: A language and environment for statistical computing.* R Foundation for Statistical Computing, Vienna, Austria. 3-900051-07-0.

Index